11/10/92

What to Solve?

D0139594

What to Solve?
Problems and Suggestions for Young Mathematicians

JUDITA COFMAN

CLARENDON PRESS · OXFORD

Oxford University Press, Walton Street, Oxford OX2 6DP

Oxford New York Toronto
Delhi Bombay Calcutta Madras Karachi
Petaling Jaya Singapore Hong Kong Tokyo
Nairobi Dar es Salaam Cape Town
Melbourne Auckland
and associated companies in
Berlin Ibadan

Oxford is a trade mark of Oxford University Press

Published in the United States
by Oxford University Press, New York

© *Judita Cofman, 1990*

First published 1990
Reprinted 1990, 1991 (twice)

British Library Cataloguing in Publication Data
Cofman, Judita
What to solve?
1. Mathematics. Problem solving
I. Title
510
ISBN 0–19–853294–6

Library of Congress Cataloging in Publication Data
Cofman, Judita.
What to solve?: problems and suggestions for young mathematicians
/Judita Cofman.
p.   cm.
Includes bibliographical references.
1. Problem solving.   2. Mathematics—Problems, exercises, etc.
I. Title.
QA63.C64   1989   510'.76—dc20
89–22150
ISBN 0–19–853294–6

Printed and bound in Great Britain by
Biddles Ltd, Guildford and King's Lynn

Preface

What to Solve? is a collection of mathematical problems for secondary school pupils interested in the subject.

There are numerous problem collections on the market in different parts of the world, many of them excellent. Is there a need for one more collection of problems? Perhaps not, but I felt tempted to convey some of my views and suggestions on learning about mathematics through problem solving. To discuss mathematical ideas without presenting a variety of problems seemed pointless. Since I had already gathered a fair number of problems from 'problem seminars' at international camps for young mathematicians which I have run in the past years, I decided to compile a book of problems and solutions from the camps. The arrangement of the text, the selection and grouping of the questions, the comments, references to related mathematical topics, and hints for further reading should illustrate my ideas about studying — or educating — through problem solving.

At the camps the age of the participants ranged from 13 to 19 and their mathematical backgrounds were tremendously varied. Consequently problems at the problem seminars varied in their degrees of difficulty. The same applies to the problems of this collection. It is hoped that any seriously interested reader, aged 13 and above, might benefit from 'bits and pieces' of the text.

The composition of the book reflects the philosophy of the problem seminars at the camps. The campers were led step by step through four stages of problem solving:

Stage 1: Encouraging independent investigation
Finding an answer to a question by one's own means brings about a pleasure known to problem solvers of all ages. We recommended that all participants, especially beginners, first attempted all the investigations of easier problems which could be tackled alone, without hints and guidance.

Stage 2: Demonstrating approaches to problem solving
Having tested the joy of independent discovery, the youngsters became more critical towards their own achievements and wondered: 'Is there a better (that is, quicker, or more elementary, or more elegant) solution to this problem?' By this time the campers were motivated, and were given the opportunity to learn

more about techniques for problem solving.

Stage 3: *Discussing solutions of famous problems from past centuries*

Detecting problems and attempting their solution has been the life-time occupation of professional mathematicians throughout the centuries. Advanced problem solvers were encouraged to study famous problems, their role in the development of mathematics and their solutions by celebrated thinkers. As well as improving their problem-solving skills they learnt to appreciate mathematics as part of our culture.

Stage 4: *Describing questions considered by eminent contemporary mathematicians*

Extensive study of research problems of modern mathematics is, generally, not possible at pre-university level. Nevertheless there are a number of questions, e.g. in number theory, geometry, or modern combinatorics, which can be understood without much previous knowledge of 'higher mathematics'. The aim of the last stage at the seminars was to describe a selection of questions which have attracted the attention of eminent twentieth-century mathematicians.

Following this pattern of four stages of problem solving, the treatment of the problems in this book is divided into four chapters. These bear the names and are guided by the ideas of the corresponding stages at the seminars. Each chapter consists of two parts, preceded by an introduction. Part I presents the problems, Part II contains solutions. The introduction describes the aims of the chapter, sketches methods for solving the problems and highlights easier problems or problems of special interest. There are three appendices at the end of the book: *Appendix I* is more than a glossary: it contains definitions and explanations about mathematical notions encountered in the book. *Appendix II* consists of biographical notes about mathematicians referred to in the text. *Appendix III* presents an extensive list of recommended further reading (including references).

Suggestions to readers: Who could use the book and how?

- *Beginners* (aged 13 +) are advised to look out for easier questions marked 'E', and for hints in the introductions to the chapters.
- *Mathematics teachers* might adapt easier questions (or even some of the harder ones) for investigations in mathematics clubs or in the classroom.
- Harder problems, aimed at *advanced pupils*, should assist not only in developing their skills at answering questions, but also in firing their curiosity and interest in further reading.

- *Competitors at mathematical Olympiads and similar contests* may find the book useful during their training. Discussion of past IMO questions is avoided (with a few exceptions); on the other hand, the book contains a number of challenging problems from the Soviet magazine *Kvant*, the German collection *Bundeswettbewerb Mathematik* and the Hungarian journal *Matematikai Lapok*.
- Last but not least, Appendix III, which contains titles for further reading, should be consulted by *advanced pupils* and *mathematics teachers*.

A few remarks about the choice of the problems and the nature of solutions

The problems in Chapters III and IV are, naturally, well known, and can be found elsewhere in the literature of mathematics. The same applies, largely, to the problems in Chapters I and II: they are well known. Although participants at the camps regularly spent long hours on projects of their own choice, often pursuing unorthodox tasks, it seems appropriate to discuss here those questions from the problem seminars which proved to be useful and enjoyable for generations of enthusiastic youngsters. The popularity of the problems vouches for their sound educational value.

The solutions given in Chapters III and IV follow, more or less, the ideas of celebrated mathematicians who attempted the problems. In Chapters I and II, however, solutions are based on answers and thought processes of typical campers. In other words, they usually differ from 'model solutions'. One can argue whether this is a proper attitude in educating. I find it not only an admissible, but also an effective, way of assisting a large group of youngsters of diverse ages and mathematical backgrounds in their attempts to do work outside the school syllabus. This approach avoids frightening beginners by too much 'perfection' and the advanced soon discovered ways of looking for 'better and more'. Readers of this book are encouraged to do the same!

London
August 1989   J.C.

Acknowledgements

I am deeply indebted to Dr Peter Neumann and to Dr Abe Shenitzer for their continuous encouragement and invaluable advice during the preparation of this book. They suggested many improvements and corrected many errors; it is impossible to thank them adequately.

I am most grateful to Terry Heard; his thorough, helpful criticism of the manuscript helped to eliminate a number of obscurities.

My thanks go to Helen Hodgson for the beautiful typing, and to Samantha Rawle for the careful proofreading.

Sincere thanks are due also to my colleagues at Putney High School, London. Without their sympathetic, friendly support the book would never have been finished. The preliminary work on the manuscript was begun during my stay at St Hilda's College, Oxford, where I enjoyed a teacher fellowship in Trinity term of the academic year 1985–86, for which I am very grateful.

I would like to use this opportunity to express my gratitude to all those involved in organizing the Mathematics Camps, where the material contained in the text was shaped. Special thanks are due to Anna Comino James and Glyn Jarrett; without them the camps could not have existed. Above all, I am indebted to the campers themselves for their enthusiastic approach to problem solving. My warmest thanks go to Alex Selby and Andrew Smith for numerous challenging comments and stimulating discussions on a wide range of mathematical topics.

Finally, I wish to thank Oxford University Press for publishing the book. I am indebted to Anthony Watkinson and to Martin Gilchrist for their support and collaboration.

Contents

Contents xiii

I Problems for investigation

Introduction

Even at a very early age, investigating a pattern of numbers or shapes can lead to intriguing discoveries (no matter how small) and may raise a number of challenging questions. Therefore investigative work seems to provide a suitable introduction to the art of problem solving. What should one investigate? And how?

This chapter contains a selection of problems, based on some popular types of investigation, such as:

1. iterating a certain procedure and analysing the results;
2. search for patterns;
3. looking for exceptions, or special cases in a pattern;
4. generalizing given problems;
5. studying converse problems.

Each of the above approaches is treated in a separate section. Here we shall explain briefly what they mean.

1. *Iterating a certain procedure* means to perform the procedure on an object (e.g. on a number, or on a shape) repeatedly; to *analyse* the result means to find out what happened. For example:

Let T_0 be an equilateral triangle of unit area. Divide T_0 into 4 equal equilateral triangles T_1 by joining the midpoints of the sides of T_0. Now remove the central triangle (Fig. 1.1). Treat the remaining 3 triangles in the same way and repeat the process — that is, iterate — n times. Find out:
(a) What is the sum S_n of the areas of the removed triangles after n steps?
(b) What happens to S_n as n tends to infinity?

In the above example iteration leads to

1 removed triangle T_1 of area $\frac{1}{4}$,

3 removed triangles T_2 of area $(\frac{1}{4})^2$,

3^2 removed triangles T_3 of area $(\frac{1}{4})^3$,

3^3 removed triangles T_4 of area $(\frac{1}{4})^4$,

1

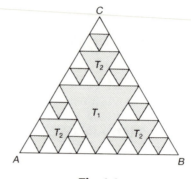

Fig. 1.1

and at the nth step to

$$3^{n-1} \text{ removed triangles } T_n \text{ of area } (\tfrac{1}{4})^n.$$

Thus:
(a) $S_n = 1 \cdot \tfrac{1}{4} + 3 \cdot (\tfrac{1}{4})^2 + 3^2(\tfrac{1}{4})^3 + \cdots + 3^{n-1}(\tfrac{1}{4})^n$

$$= \tfrac{1}{4} \, \frac{1 - (\tfrac{3}{4})^n}{1 - \tfrac{3}{4}} = 1 - (\tfrac{3}{4})^n.$$

(b) As n tends to infinity, $(\tfrac{3}{4})^n$ tends to 0, hence S_n tends to 1.

It is recommended that younger readers start Section 1 with Problems 1(E) and 5(E).

   2. *Search for patterns* has never ceased to fascinate people throughout the ages.

   For example, the ancient Greeks considered odd numbers superior to even numbers, because their sums

$$1, \; 1+3, \; 1+3+5, \; 1+3+5+7$$

could be represented by sets of dots neatly fitting into squares (Fig. 1.2).

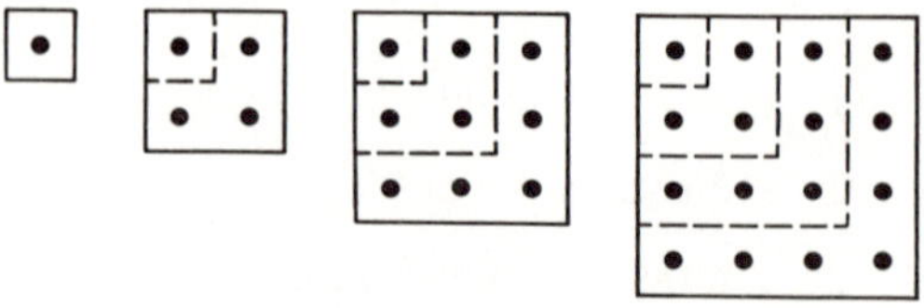

Fig. 1.2

The problems in Section 2 may stimulate readers to search for patterns 'of their own'. Beginners could benefit from studying number patterns on the chessboard (Problem 11(E)), divisions of regular polygons into rhombuses (Problem 15(E)), or 'Fibonacci squares' (Problem 16(E)). Advanced problem solvers might enjoy Problems 18–20.

3. *Looking for exceptions and special cases* is important for various reasons:

(a) In practical situations exceptional cases often provide the best possible solutions (e.g. the shortest path between two points).

(b) Overlooking exceptions in problem solving might lead to errors.

This is illustrated by the following problem:

Figure 1.3(a) shows three mirrors of length ℓ forming a triangle ABC. A light source is placed at a point S of AB, at a distance d from A. A light ray, emerging from S at an angle of 60° with SB, gets reflected from the sides of triangle ABC until it returns to S. Find the length of the light ray's path in terms of ℓ.

It is easy to prove that the length of the light ray's path, as shown in Fig. 1.3(a), is 3ℓ. This particular answer does not involve the distance d of

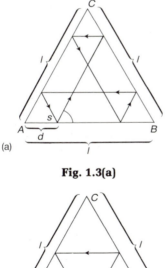

(a)

Fig. 1.3(a)

(b)

Fig. 1.3(b)

S from A; therefore one may jump to the conclusion that the length of the light ray's path is the same for all positions of S on AB. In fact, the conclusion is true — with the exception of one single case: If S is at the midpoint of AB, the length of the path is $\frac{3}{2}\ell$ (Fig. 1.3(b)).

It is worthwhile noticing that special cases often represent extremal solutions to problems (i.e. maxima or minima). The quickest, and most convenient, way of detecting them is usually by using calculus. Nevertheless, it always remains a challenge for problem solvers to work with methods as elementary as possible. Problem 26 is a good example of a question which can be easily tackled by differentiation, and at the same time can be solved using geometry and trigonometry only. Readers are advised to try both approaches.

4. What does it mean *to generalize a given problem*? Suppose that a problem P involved the investigation of certain properties of a set S of objects. One might wish to study similar properties of an extended set S', containing S as a subset. This leads to a new problem P' with P as one of its special cases. We say that P' is a *generalization* of P.

A problem can have various generalizations. For example: An equilateral triangle can be viewed as a regular polygon, or as a set of points in space with equal mutual distances. Thus the problem: 'Construct an equilateral triangle of given side length a can be generalized in two ways:

(a) Construct a regular n-gon of given side length a, or

(b) Construct n points in space with mutual distances equal to a.

Apart from the problems presented in Section 4, readers are encouraged to generalize (and solve) some of the problems from previous sections.

5. In the following example Theorem[1] B is the *converse* of Theorem A:

Theorem A: Triangle ABC has two equal sides;
                therefore
                triangle ABC has two equal angles.

Theorem B: Triangle ABC has two equal angles;
                therefore
                triangle ABC has two equal sides.

There are theorems whose converses are false. (For example, the Theorem 'If x is a positive number, then x^2 is positive' is valid, but its converse 'If x^2 is positive, then x must be positive' is false.)

After proving a particular theorem one should attempt to formulate its converse, and to investigate whether this converse is true or false. (A theorem can have more than one converse.)

Theorems whose converses are also true are especially important. They lead to conditions which are *necessary* and at the same time *sufficient* for the validity of certain properties. For example, Theorems A and B imply that for

---

[1] For the definition of a theorem see p. 145.

a triangle *ABC* to be isosceles (that is, to have two equal sides) it is necessary as well as sufficient to have two equal angles. (*Necessary* means: $\triangle ABC$ cannot have two equal sides without having two equal angles. *Sufficient* indicates that if in $\triangle ABC$ two angles are equal, then $\triangle ABC$ must have equal sides.) In other words, the statements of Theorems *A* and *B* can be combined into

    Theorem C: A triangle *ABC* has two equal sides
                   if and only if two of its angles are equal.

Readers are recommended to study Problem 46 describing a well-known condition which is sufficient and necessary for two numbers to be relatively prime. The statement in Problem 46 will be used later on. The Theorems of Fermat (Problem 45) and Desargues (Problem 48) will also be referred to in Chapter III.

Problems 41(E), 42(E) and 44(E) are suitable for beginners.

Part I: Problems

Section 1: Iterating

Problem 1 (E)

Start with an experiment:

Take any natural number n_0, say $n_0 = 928$. Form a new number n_1 by adding the digits of n_0 ($n_1 = 9 + 2 + 8 = 19$).

Iterate the procedure of adding digits, that is add the digits of n_1 to obtain n_2, add the digits of n_2 to obtain n_3, add the digits of n_3 to obtain n_4, and so on. (If $n_1 = 19$, then $n_2 = 1 + 9 = 10$, $n_3 = 1 + 0 = 1$, $n_4 = 1$, $n_5 = 1, \ldots$).

In the above example the process of creating new numbers from n_0 has ended with n_3. This suggests two questions:

(a) Is it true for *every* natural number n_0, that the above procedure of creating new numbers $n_1, n_2, n_3, \ldots$ ends after a finite number of steps in a one-digit number n_k?

(b) If the answer to question (a) is 'yes', discover the connection between n_k and n_0. (In other words, find out for which numbers n_0 is $n_k = 1$, and for which n_0 is $n_k = 2, 3, 4, \ldots, 9$).

Problem 2 (E)

Place four non-negative integers a_0, b_0, c_0, d_0 around a circle: for example, those shown in Fig. 1.4.

For any two consecutive numbers on the circle form the absolute value of their differences:

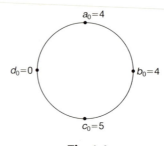

Fig. 1.4

$$a_1 = |a_0 - b_0|, \ b_1 = |b_0 - c_0|, \ c_1 = |c_0 - d_0|,$$
$$d_1 = |d_0 - a_0|$$

and put these numbers next to one another around a new circle (Fig. 1.5).

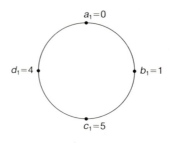

Fig. 1.5

Iterate the above process: Form the absolute values $a_2 = |a_1 - b_1|$, $b_2 = |b_1 - c_1|$, $c_2 = |c_1 - d_1|$, $d_2 = |d_1 - a_1|$ and put them around a new circle; from these numbers, in a similar way, form a circle with $a_3 = |a_2 - b_2|$, $b_3 = |b_2 - c_2|$, $c_3 = |c_2 - d_2|$, $d_3 = |d_2 - a_2|$, and so on.

In general let $a_{i+1} = |a_i - b_i|$, $b_{i+1} = |b_i - c_i|$, $c_{i+1} = |c_i - d_i|$, and $d_{i+1} = |d_i - a_i|$.

The problem is to find out whether the process of creating new circles ends after some steps, and, if so, how. In the example when $a_0 = 4$, $b_0 = 4$, $c_0 = 5$, $d_0 = 0$, the numbers at the fourth step a_4, b_4, c_4, d_4 all become 0s and after that step no new circle can be formed.

Is it true in general that for *any* choice of four non-negative integers a_0, b_0, c_0, d_0 the above procedure leads to a circle with four 0s?

Problem 3 (E)

There are many ways of constructing new triangles from a given one; we shall investigate the following problem:

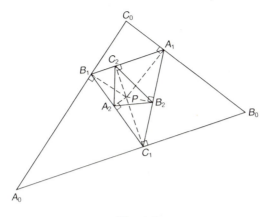

Fig. 1.6

Start from a triangle $A_0B_0C_0$. Choose an arbitrary point P inside the triangle and from P drop a perpendicular onto each side of $\Delta A_0B_0C_0$: PA_1 onto B_0C_0, PB_1 onto A_0C_0 and PC_1 onto A_0B_0. Join the points A_1, B_1, C_1 to obtain $\Delta A_1B_1C_1$. Iterate the procedure: from P drop perpendiculars PA_2, PB_2, PC_2 onto the sides B_1C_1, C_1A_1 and A_1B_1 of $\Delta A_1B_1C_1$ and join A_2, B_2 and C_2 to form $\Delta A_2B_2C_2$. In the same way, by using P and $\Delta A_2B_2C_2$ construct $\Delta A_3B_3C_3$, and so on.

Figure 1.6 suggests that the process of obtaining new triangles, all of different sizes, never ends. Nevertheless, we can ask:

Is one of the triangles $A_1B_1C_1$, $A_2B_2C_2$, $A_3B_3C_3$, ... similar to the original triangle $A_0B_0C_0$?

Problem 4

Iterated geometric constructions can lead to interesting number series (Fig. 1.7).

Figure 1.7 shows two circles a and b of radius 1 touching at a point P. t is a common tangent, touching a at Z and b at Z_1. c_1 is the circle touching a, b and t; c_2 is the circle touching a, b and c_1; c_3 is the circle touching a, b and c_2.

The procedure of constructing circles touching a, b and c_k, $k = 1, 2, 3$, ..., can be continued endlessly. The diameters d_1, d_2, ... of the circles c_1, c_2, ... form an infinite number sequence. The sum $s_k = d_1 + d_2 + \cdots + d_k$ gets larger as k increases ($d_1 < d_1 + d_2 < d_1 + d_2 + d_3 < \cdots$). Clearly, s_k never becomes greater than PX_1, the distance of P from t, which is 1; on the other hand s_k gets as close to PX_1 as we like.

Thus the sum s_k tends to 1 as k tends to infinity. The problem is:

Find the terms d_1, d_2, d_3, ..., d_k of the number series

$$d_1 + d_2 + d_3 + \cdots = 1.$$

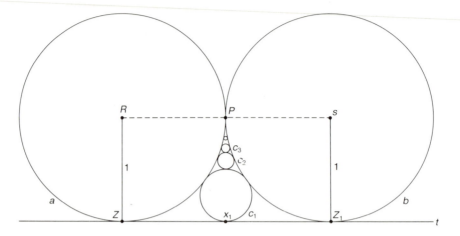

Fig. 1.7

Remarks: The numbers $d_1, d_2, d_3, \ldots$ and their sum 1 are contained in the harmonic triangle, a number pattern investigated by Leibniz (see Appendix I).

Problem 5 (E)

In this problem we explore a connection between the Fibonacci numbers and a sequence of pentagons and pentagrams.

Recall that the Fibonacci numbers are the elements of the sequence 1, 1, 2, 3, 5, 8, 13, 21, . . . The kth Fibonacci number f_k is the sum of the two preceding elements of the sequence:

$$f_k = f_{k-1} + f_{k-2}.$$

We shall construct a sequence with a similar property:

Start with a regular pentagon P_1. Extend its sides to obtain a pentagram (that is, a five-pointed star) S_1. Join the vertices of S_1 to obtain a new regular pentagon P_2, and extend the sides of P_2 to form a pentagram S_2 (Fig. 1.8). This process can be iterated to obtain an infinite sequence of shapes: $P_1, S_1, P_2, S_2, P_3, S_3, \ldots$.

Denote the side lengths of $P_1, P_2, P_3, \ldots$ by $a_1(= a), a_2, a_3, \ldots$ and the side lengths of $S_1, S_2, S_3, \ldots$ by $b_1(= b), b_2, b_3, \ldots$ respectively. Our aim is to study the number sequence S:

$$a, b, a_2, b_2, a_3, b_3, \ldots.$$

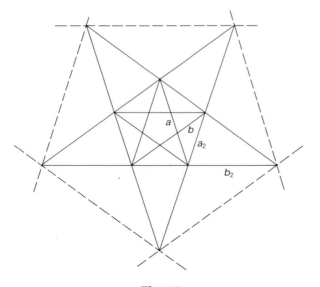

Fig. 1.8

(a) Prove that each term of S is the sum of the two preceding terms.
(b) Show that for any $k = 3, 4, 5, \ldots$ the kth term of S can be written in the form $f_{k-2}a + f_{k-1}b$, where f_{k-2} and f_{k-1} are the $(k-2)$nd and the $(k-1)$st Fibonacci numbers.

Problem 6 (E) *(Kvant, M871, 1984, No. 10)*

The numbers $a_1, \ldots, a_9$ in the cells of the square $ABCD$ are all either $+1$ or -1. From $ABCD$ a new square $A_1B_1C_1D_1$ is constructed as shown in Fig. 1.9: each number in square ABCD is replaced by the product of the

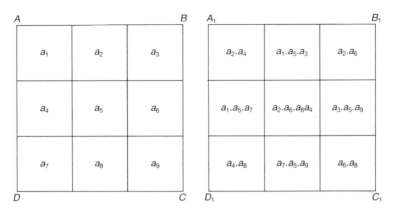

Fig. 1.9

numbers in the neighbouring cells. (Two cells are called neighbouring if they have an edge in common.)

Iterate this process and find out what happens.

Problem 7

Start with the number $a_1 = 7$. Form its seventh power: $a_2 = a_1^7$. Iterate the process of forming seventh powers:

$$a_3 = a_2^7, \ a_4 = a_3^7, \ \ldots \ a_k = (a_{k-1})^7.$$

Investigate: How does the final digit of a_k depend on k?

Problem 8

Find out: Is there any prime number in the infinite sequence

$$10001, \ 100010001, \ 1000100010001, \ \ldots \ ?$$

Problem 9 (*Kvant*, M904, 1985, No. 5)

Let A be a number of the form

$$A = 10^n a_n + 10^{n-1} a_{n-1} + \cdots + 10 a_1 + a_0,$$

where $a_0, a_1, \ldots, a_n$ are integers between 0 and 9 inclusive.

Obtain the number A_1 from A according to the rule:

$$A_1 = D(A) = a_n + 2a_{n-1} + 2^2 a_{n-2} + \cdots + 2^{n-1} a_1 + 2^n a_0.$$

Iterate this construction to obtain

$$A_2 = D(A_1), \ A_3 = D(A_2), \ \ldots$$

(a) Prove that for any natural number A the above process leads to a number $A_k < 20$ such that $D(A_k) = A_k$.

(b) If $a = 19^{85}$, consider the sequence $A_1 = D(A)$, $A_2 = D(A_1)$, $\ldots$ Determine the number A_k for which $D(A_k) = A_k$.

Problem 10 (*Bundeswettbewerb*, 1978, 1st round; [90])

The moves of a knight on the chessboard are modified: In each move the knight traverses p squares horizontally and q vertically, or q squares horizontally and p vertically, where p and q are arbitrary, given positive integers. The chessboard is unlimited.

If, after n steps, the knight returns to its starting position, is it true that n must be an even number?

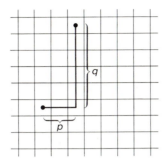

Fig. 1.10

Section 2: Search for patterns

We start with some number patterns obtained from the chessboard.

Problem 11 (E)

(a) A 'chessboard' is bounded from the top and from the left only (Fig. 1.11). A rook is placed on the square A in the upper left corner, and can move 'horizontally' or 'vertically'.

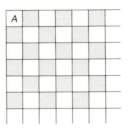

Fig. 1.11

For each square on Fig. 1.11 find the number of shortest paths the rook can take from A to that square, and write this number in the square.

Investigate the resulting number pattern.

(b) Instead of a rook put a king on the square A in Fig. 1.11. The king is allowed to move in three directions only: from 'left to right' ($\rightarrow$) 'vertically downwards' ($\downarrow$) and 'diagonally downwards to the right' ($\searrow$). For each square in Fig. 1.11 find the number of paths leading the king from A to that square, and write the number in the square.

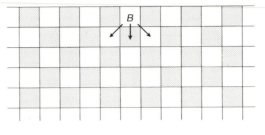

Fig. 1.12

Investigate the number pattern.

(c) The chessboard in Fig. 1.12 is bounded from the top only. A king, placed on square *B*, is allowed to move in three directions: 'diagonally downwards to the left', 'vertically downwards' and 'diagonally downwards to the right'. Construct a number pattern on the board by writing into each square the number of paths leading the king from *B* to that square.

Investigate this number pattern.

The examples in Problem 12 show number patterns obtained from arithmetic progressions:

Problem 12 (parts (a) and (b) are not difficult)

(a) (E) In Fig. 1.13 calculate

(i) the sum of the numbers in each square:

$$s_1 = 1, s_2 = 1 + 2 + 2 + 4 = 9,$$

$$s_3 = 1 + 2 + 3 + 2 + 4 + 6 + 3 + 6 + 9 = ?$$

$$s_n = ?$$

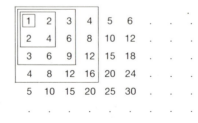

Fig. 1.13

and

(ii) the sum of the numbers in each 'corridor' between the two squares:

$$c_1 = 2 + 4 + 2 = 8, c_2 = 3 + 6 + 9 + 6 + 3 = ?$$

$$c_n = ?$$

What can be learnt from the pattern of the numbers s_n and c_n?

  (b) (E) Investigate the following number array:

1	3	5	7	9	11	13	.
1	4	7	10	13	16	19	.
1	5	9	13	17	21	25	.
1	6	11	16	21	26	31	.
1	7	13	19	25	31	37	.
1	8	15	22	29	36	43	.

Fig. 1.14

(c) The number array below is known as Sundaram's sieve:

4	7	10	13	16	19	22	25	.
7	12	17	22	27	32	37	42	.
10	17	24	31	38	45	52	59	.
13	22	31	40	49	58	67	76	.

Fig. 1.15

Prove that

 (i) if k is any number in the sieve, then $2k + 1$ is not a prime number, and, conversely,

(ii) if $2k + 1$ is not a prime number, then k is in the sieve.

  (d) Use the number array in Fig. 1.16 to evaluate the sum of the squares $S_{2,n} = 1^2 + 2^2 + \cdots + n^2$.

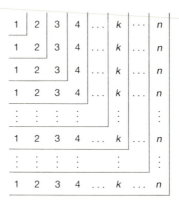

Fig. 1.16

Problem 13 (E) (*Kvant*, M436, 1978, No. 1)

$a_1, a_2, a_3, \ldots, a_{10}, b_1, b_2, \ldots, b_{10}$ are twenty numbers. The sums $a_1 + b_1$, $a_1 + b_2, \ldots, a_1 + b_{10}, a_2 + b_1, a_2 + b_2, \ldots, a_2 + b_{10} \ldots, a_{10} + b_1, a_{10} + b_2$, $\ldots a_{10} + b_{10}$ are 100 numbers, not necessarily distinct.

Prove that it is possible to arrange the hundred sums in ten groups of ten numbers in each such that the sums of the numbers in all groups are the same.

Problem 14 (E)

A triangle with angles 36°, 72°, 72° is isosceles, and it can be divided into two triangles, each of which is again isosceles (Fig. 1.17).

Find all types of isosceles triangles which can be divided into two isosceles triangles.

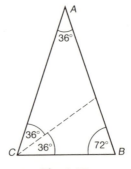

Fig. 1.17

Problem 15 (E)

A regular hexagon can be divided into three rhombuses, and a regular octagon can be divided into six rhombuses, as shown in Fig. 1.18.

Is it possible to divide any regular polygon with an even number of sides into rhombuses?

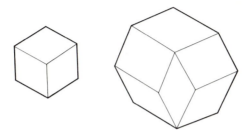

Fig. 1.18

Problem 16 (E)

The side lengths of the squares in Fig. 1.19 are the Fibonacci numbers 1, 1, 2, 3, 5, 8, ... The figure indicates that the centres of these squares lie on two mutually perpendicular straight lines. Is this true?

Problem 17

(a) Find the number of those solutions of the equation $x + y + z + w = 12$ which are positive integers.
(b) Express the number of the positive integer solutions of the equation

$$x_1 + x_2 + \cdots + x_n = k$$

in terms of n and k.

Problem 18

The numbers $2 + 1$, $2^2 + 1$ are relatively prime[1], and so are the numbers $2 + 1$, $2^4 + 1$ and $2^2 + 1$, $2^4 + 1$.

Is this true in general? That is, are any two members of the sequence

$$2 + 1, 2^2 + 1, 2^4 + 1, 2^8 + 1, \ldots$$

relatively prime?

---

[1] See Appendix I.

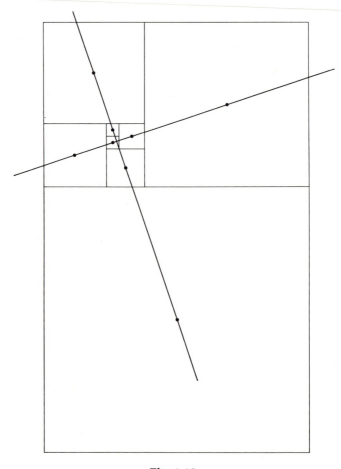

Fig. 1.19

Problem 19 (*Bundeswettbewerb Mathematik*, 1976 1st round [90])

Let a/b be a fraction in its simplest terms, that is, the greatest common divisor of a and b is 1. In the tree diagram shown in Fig. 1.20 the fraction a/b has two successors:

$$\frac{a}{a+b} \quad \text{and} \quad \frac{b}{a+b}.$$

Each of these successors has two successors of its own, constructed in the same way (the successors of $a/(a + b)$ are

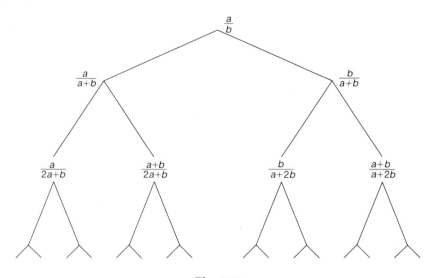

Fig. 1.20

$$\frac{a}{2a + b} \quad \text{and} \quad \frac{a + b}{2a + b},$$

and the successors of $b/(a + b)$ are

$$\frac{b}{a + 2b} \quad \text{and} \quad \frac{a + b}{a + 2b}.$$

Find a and b such that the tree diagram starting with a/b consists of all positive fractions less than 1.

Problem 20 (MU [94])

Peter and Paul play the following game: Starting with Peter they take turns in calling out the greatest odd divisor of the consecutive natural numbers 1, 2, 3, 4, 5, 6, ... If the divisor called out is of the form $4k + 1$, then Peter pays £1 (in the original version 1 DM) to Paul, otherwise Paul pays £1 to Peter. After a while the players stop.

Can you predict who will have won?

Section 3: Exceptions and special cases

Problem 21 (E)

Let ABC be a right-angled triangle with hypotenuse AB, and let M be an

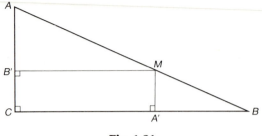

Fig. 1.21

arbitrary point on *AB*. By dropping perpendiculars *MA'* onto *BC* and *MB'* onto *AC* a rectangle *MA'CB'* is constructed (Fig. 1.21).

Any *isosceles* right-angled triangle has the remarkable property that all rectangles inscribed in it in the above way have the same perimeter equal to *AC* + *CB*. Hence the problem:

'Inscribe into △*ABC* a rectangle *MA'CB'* of given perimeter 2*s* such that 2*s* ≠ *AC* + *CB*'

has no solution if *AC* = *CB*.

Investigate: Is the isosceles right-angled triangle the only exception? Can one inscribe rectangles of given perimeter 2*s* in right-angled triangles *ABC*, such that 2*s* ≠ *AC* + *CB*, in the way shown in Fig. 1.21, in any other case?

Problem 22 (E)

(a) Are there any convex polygons other than obtuse-angled triangles in which one angle is greater than the sum of the remaining angles?
(b) Are there convex *n*-gons with *n* acute angles for any value of *n*?

Problem 23 (E)

Prove that in any parallelogram the bisectors of the angles determine a rectangle *R* (Fig. 1.22). When is *R* a square?

Problem 24

A cat knows the three exits of a mouse hole. Where should the cat sit so that its distance from the furthest exit is a minimum?

Problem 25 (*Matem. Lapok*, 985, 1960, No. 3–4)

For which values of the real number *a* does the equation

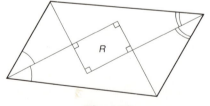

Fig. 1.22

$$[\sqrt{(a + \sqrt{(a^2 - 1)})}]^x + [\sqrt{(a - \sqrt{(a^2 - 1)})}]^x = 2a$$

have real solutions in x?

Discuss the number of solutions.

Problem 26

Among all quadrilaterals $ABCD$ with given side lengths $AB = a$, $BC = b$, $CD = c$, and $DA = d$ find the one with the greatest area.

Problem 27 (Kvant, M892, 1985, No. 3)

(a) Prove that there are infinitely many square numbers of the form $2^m + 2^k$, where m and k are distinct positive integers.
(b) Do the same for $3^m + 3^k$.
(c) Investigate: Are there infinitely many square numbers ámong the numbers $4^m + 4^k$, $5^m + 5^k$, $6^m + 6^k$, $7^m + 7^k$, where m and k are distinct, positive integers?

Problem 28 (Kvant, M920, 1985, No. 8)

(a) Find at least one solution of the equation

$$x^3 + y^3 + z^3 = x^2y^2z^2$$

in natural numbers x, y, z.
(b) Are there any values of the natural numbers $n \neq 1$ such that the equation

$$x^3 + y^3 + z^3 = nx^2y^2z^2$$

has solutions in natural numbers?

Problem 29 (Bundeswettbewerb, 1977, 1st round; [90])

The number 50 is expressed as the sum of some natural numbers, not necessarily different from one another. The product of the summands is divisible by 100. What is the greatest possible value of the product?

Problem 30

P is a polyhedron whose edges are all of the same length, and all touch a given sphere S. Is it always possible to construct a sphere S' passing through all vertices of P?

Section 4: Generalizing given problems

Problem 31 (E) (from the collection of Pappus)

ABC is an arbitrary triangle and $ABDE$ and $CBGF$ are arbitrary parallelograms constructed on two of the sides. Extend ED and FG to meet in H, and

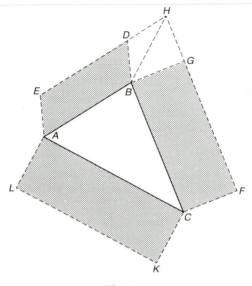

Fig. 1.23

construct a parallelogram $ACKL$ on the third side AC such that AL and CK are equal to HB and parallel to it (Fig. 1.23).

Prove that the area of $ACKL$ is the sum of the areas of $ABDE$ and $CBGF$.

Problem 32 (E)

$OABC$ is a pyramid whose edges OA, OB and OC all meet at right angles. Prove that (Area $AOB)^2$ + (Area $BOC)^2$ + (Area $COA)^2$ = (Area $ABC)^2$.

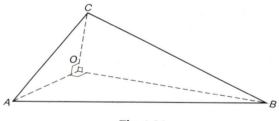

Fig. 1.24

Problem 33

(a) Prove that the product of two positive numbers a, b whose sum is constant attains its maximal value when $a = b$.

(b) Generalize the statement in (a) for n positive numbers, where n is any natural number.

Problem 34

(a) If $A_1, A_2, \ldots, A_n$ are n points on a circle of radius 1, prove that the sum of the squares of their mutual distances is not greater than n^2.
(b) Does the same statement hold for points on a sphere of radius 1?

Problem 35

(a) Prove that the set of points in the plane whose squares of distances from two given points A_1 and A_2 add up to the same constant is either a circle, a point, or the empty set.
(b) Investigate the general problem: If $A_1, A_2, \ldots, A_n$ are given points in the plane, find out what is the nature of the set of points P for which

$$\alpha_1|PA_1|^2 + \alpha_2|PA_2|^2 + \cdots + \alpha_n|PA_n|^2 = \beta,$$

where $\alpha_1, \alpha_2, \ldots, \alpha_n$ and β are given numbers and $|PA_i|$ is the distance of P from A_i.

Problem 36

Generalize the problem of finding Pythagorean triples as follows: Show that the equation

$$x_1^2 + x_2^2 + x_3^2 + \cdots + x_{n-1}^2 + x_n^2 = y_n^2$$

has non-zero integer solutions $x_i = a_i, i = 1, 2 \ldots, n, y_n = b_n$ for all $n = 2, 3, 4, \ldots$.

Problem 37

(a) Given n straight lines in the plane, find the greatest possible number of regions into which the lines can divide the plane.
(b) How could the above problem be generalized in space?

Problem 38

A Latin square of order n is a matrix constructed from n symbols as entries, such that each symbol appears in every row and in every column exactly once. For example:

$$M_1 = (1), \quad M_2 = \begin{pmatrix} 0 & 1 \\ 1 & 0 \end{pmatrix}, \quad M_3 = \begin{pmatrix} 0 & 1 & 2 \\ 1 & 2 & 0 \\ 2 & 0 & 1 \end{pmatrix}, \quad M_4 = \begin{pmatrix} 0 & 1 & 2 & 3 \\ 1 & 2 & 3 & 0 \\ 2 & 3 & 0 & 1 \\ 3 & 0 & 1 & 2 \end{pmatrix}.$$

In the above examples, when in the square M_i^2 of the matrix M_i the entries are replaced by their remainders modulo i, the resulting matrix M_i^* is of the form

$$M_1^* = (1), \quad M_2^* = \begin{pmatrix} 0 & 1 \\ 1 & 0 \end{pmatrix}, \quad M_3^* = \begin{pmatrix} 2 & 2 & 2 \\ 2 & 2 & 2 \\ 2 & 2 & 2 \end{pmatrix}, \quad M_4^* = \begin{pmatrix} 2 & 0 & 2 & 0 \\ 0 & 2 & 0 & 2 \\ 2 & 0 & 2 & 0 \\ 0 & 2 & 0 & 2 \end{pmatrix}.$$

Investigate whether this pattern continues. Is is true that if M_n is a Latin square of the form

$$M_n = \begin{pmatrix} 0 & 1 & 2 & . & . & . & n-1 \\ 1 & 2 & 3 & . & . & . & 0 \\ 2 & 3 & 4 & . & . & . & 1 \\ . & . & . & & & & . \\ . & . & . & & & & . \\ . & . & . & & & & . \\ n-1 & 0 & 1 & . & . & . & n-2 \end{pmatrix}$$

then the corresponding matrix M_n^* contains one type of symbols for n odd, and two types of symbols for n even?

Finally, two problems on generalized chessboards:

Problem 39 (Soviet Olympiad 1971)

What is the smallest number of rooks governing the $n \times n \times n$ chessboard?

Problem 40

A 'three-dimensional chessboard' is constructed from congruent cubes arranged in layers. The board is limited by three planes only: from 'below', from 'behind' and from the 'left' (Fig. 1.25). A rook, placed at O, is allowed to move along lines parallel to the x, y and z axes.

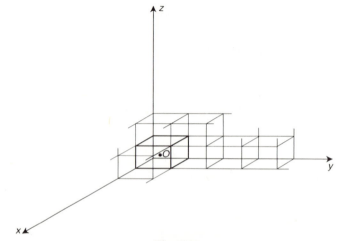

Fig. 1.25

(a) Find the number of shortest routes leading from O to any cell (that is cube) of the board
(b) Examine the number pattern obtained.

Section 5: Converse problems

Problem 41 (E)

In a triangle ABC the bisectors of the angles at A and B meet the opposite sides in points A' and B' respectively. O is the common point of the bisectors (Fig. 1.26).
(a) Prove that: If the triangle ABC is isosceles, with $\angle CAB = \angle CBA$, then $OB' = OA'$.
(b) Is the converse true: If $OB' = OA'$ does it follow that $\angle CAB = \angle CBA$?

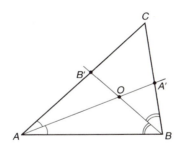

Fig. 1.26

Problem 42 (E)

If P is a point inside a square $ABCD$, then the sum of its distances from the straight lines AB and DC is equal to the sum of its distances from the straight lines DA and CB.

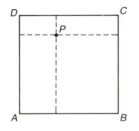

Fig. 1.27

Investigate the converse problem:

Given that P is a point in the plane of $ABCD$ such that the sum of the distances of P from AB and DC is equal to the sum of the distances of P from AD and BC, find the locus of P.

Solve the same problem when $ABCD$ is a rectangle with $AB > CD$.

Problem 43

Figure 1.28 shows two pentagons: $P = A_1A_2A_3A_4A_5$ and $P' = M_1M_2M_3M_4M_5$. The vertices of P' are the midpoints of the sides of P.

If P is given, the construction of P' is straightforward. Study the converse problem:

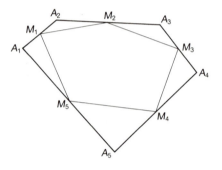

Fig. 1.28

(a) Construct P if P' is given.
(b) Investigate the case when P' is an n-gon for any natural number n.

Problem 44 (E) (Plutarch)

(a) Prove that if t is a triangular number then $8t + 1$ is a square number.
(b) If $8t + 1$ is a square number, is t necessarily a triangular number?

Problem 45

(a) Prove the statement known as the Little Theorem of Fermat: If p is a prime, and a is any natural number, then $a^p - a$ is divisible by p.
(b) The statement (S): 'If n is a prime number then n divides $2^n - 2$' is a special case of Fermat's Little Theorem, when $a = 2$.
    Prove that the converse of (S) does not hold by verifying that 341 is a composite number which divides $2^{341} - 2$.
(c) Show that there are infinitely many composite numbers n dividing $2^n - 2$.

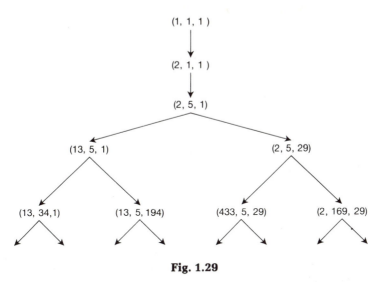

Fig. 1.29

Problem 46

Prove that two natural numbers x and y are relatively prime[1] if and only if there exist two integers a and b such that $ax + by = 1$.

Problem 47

(a) The triples of the tree diagram in Fig. 1.29 are solutions (x, y, z) of the equation

$$x^2 + y^2 + z^2 = 3xyz.$$

Explain the method for constructing the tree diagram.

(b) Is the converse true? If a, b, c are solutions of the equation $x^2 + y^2 + z^2 = 3xyz$ in natural numbers, does one of the triples (a, b, c), (a, c, b), (b, a, c), (b, c, a), (c, a, b), (c, b, a) belong to the tree diagram?

Problem 48

(a) Prove the theorem of Desargues in space:
If $A_1B_1C_1$ and $A_2B_2C_2$ are two triangles, not in the same plane, such that the straight lines A_1A_2, B_1B_2 and C_1C_2 meet in a common point S and if the straight lines A_1B_1 and A_2B_2 meet in P, B_1C_1 and B_2C_2 meet in Q, and C_1A_1 and C_2A_2 meet in R, then P, Q and R are collinear (Fig. 1.30).

(b) Prove the converse of the above theorem.

[1] See Appendix I.

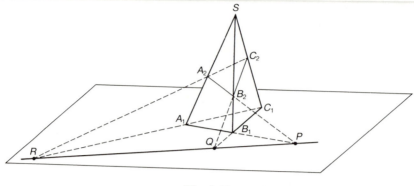

Fig. 1.30

Problem 49

(a) Prove that in a triangle with side lengths 3, 4, and 5 the radius of the inscribed circle is 1.

(b) Prove that the converse is also true: If in a triangle whose side lengths are integers, the radius of the inscribed circle is 1, then the side lengths are 3, 4, 5.

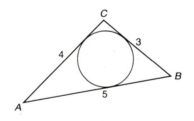

Fig. 1.31

Problem 50

(a) Prove that every plane meeting a sphere in more than one point cuts the sphere in a circle.

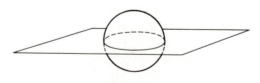

Fig. 1.32

(b) Is the converse true: If any plane meeting a surface S in more than one point intersects it in the points of a circle, is S necessarily a sphere?

Part II: Solutions

Section 1: Iterating

Problem 1 (E)

(a) Our first task is to investigate whether any natural number n_0 with more than one digit is greater than the sum of its digits. n_0 can be written in the form

$$n_0 = 10^m a_m + 10^{m-1} a_{m-1} + \cdots + 10^2 a_2 + 10 a_1 + a_0,$$

where $a_m, a_{m-1}, \ldots, a_2, a_1, a_0$ are the digits of n_0. The coefficient a_m is greater than 0, and since n_0 has more than one digit, $m \geq 1$.

Denote by n_1 the sum $a_m + a_{m-1} + \cdots + a_2 + a_1 + a_0$. From

$$a_0 = a_0$$

$$10a_1 \geq a_1$$

$$\vdots$$

$$10^{m-1} a_{m-1} \geq a_{m-1}$$

$$10^m a_m > a_m$$

it follows that

$$a_0 + 10a_1 + \cdots + 10^{m-1} a_{m-1} + 10^m a_m > a_0 + a_1 + \cdots + a_{m-1} + a_m.$$

Thus $n_0 > n_1$.

Similarly, if n_1 has more than one digit, then the sum of its digits n_2 is less than n_1; if n_2 has more than one digit, then the sum of its digits n_3 is less than n_2, and so on. However, a decreasing sequence of positive numbers cannot be continued endlessly. So, after a certain number k of steps a one-digit number n_k must be reached.

(b) The difference $d = n_0 - n_1$ can be expressed in the form

$$d = (10^m - 1)a_m + (10^{m-1} - 1)a_{m-1} + \cdots + (10^2 - 1)a_2 + (10 - 1)a_1.$$

In the above expression each summand $(10^i - 1)a_i$ is divisible by 9. This is so because

$$10^i - 1 = (10 - 1).(10^{i-1} + 10^{i-2} + \cdots + 10 + 1)$$

$$\text{for } i = 1, 2, \ldots, m.$$

It follows that d is divisible by 9. Thus n_0 and n_1, when divided by 9, leave the same remainder. Similarly n_1 and n_2, n_2 and n_3, n_3 and n_4, . . ., n_{k-1} and n_k must leave the same remainder when divided by 9.

This solves the problem:

The one-digit number n_k at the end of the sequence $n_0 > n_1 > n_2 > \cdots > n_k$ is the remainder of the division of n_0 by 9.

Problem 2 (E)

Suppose that, starting from four non-negative integers a_0, b_0, c_0, d_0, after k steps a circle carrying four zeros $a_k = b_k = c_k = d_k = 0$ was obtained. This means that in the previous step (if $k > 1$) the numbers a_{k-1}, b_{k-1}, c_{k-1}, d_{k-1} were all equal. If they were all even, then one step before that (provided $k > 2$) the numbers a_{k-2}, b_{k-2}, c_{k-2}, d_{k-2} were either all even or all odd. If a_{k-1}, b_{k-1}, c_{k-1}, d_{k-1} were all odd, then the numbers in the previous step formed two pairs (a_{k-2}, c_{k-2}) and (b_{k-2}, d_{k-2}), one consisting of even and the other of odd numbers.

This reasoning suggests the following method of solving the problem: Investigate the distribution of odd and even numbers around the circle. At the start six cases can occur:

Case 1—all numbers a, b, c, d are even;
Case 2—three numbers are even and one is odd;
Case 3—two numbers next to one another are even, the remaining two are odd;
Case 4—two numbers opposite one another are even, the remaining two are odd;
Case 5—one number is even and three are odd;
Case 6—all numbers are odd.

After the first step Case 6 is reduced to Case 1. Case 4 will be reduced to Case 1 in two steps, Case 3 in three steps and Cases 2 and 5 in four steps.

Having reached the stage when all numbers around the circle are even, one recalls that any even number is of the form $n = 2n'$, where n' is either even or odd. There are again six possibilities for the distribution of the numbers $2n'$ around the circle according to the parity of n'. Each of these cases will be reduced after at most four steps to the case when all numbers around the circle will be divisible by 4. The next steps lead to circles with all numbers divisible by 8, 16, 32, . . ., 2^t for any positive integer t.

On the other hand, the numbers on successive circles do not increase since the absolute difference of two non-negative numbers x, y cannot be greater than x or y. The only possibility for a_k, b_k, c_k and d_k to be divisible by any positive power of 2 is that all of them are 0.

Thus all circles carrying four non-negative integers will be eventually transformed into a circle with four 0s.

Problem 3 (E)

Join P to A_0, B_0 and C_0; in this way the angles of $\Delta A_0 B_0 C_0$ are divided into two parts: x, x'; y, y' and z, z' (Fig. 1.33).

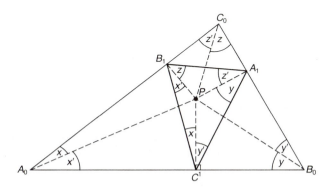

Fig. 1.33

The quadrilateral $A_0 C_1 P B_1$ is cyclic since its opposite angles at B_1 and C_1 add up to 180°. Hence, $\angle B_1 C_1 P_1 = x$ and $\angle P B_1 C_1 = x'$. The quadrilaterals $B_0 A_1 P C_1$ and $C_0 B_1 P A_1$ are also cyclic. Hence $\angle P C_1 A_1 = y'$, $\angle P A_1 C_1 = y$, $\angle P A_1 B_1 = z'$ and $\angle A_1 B_1 P = z$.

This shows that the angles of $\Delta A_1 B_1 C_1$ are formed by the pairs x, y', y, z', and z, x'. By using the same method twice more it is not difficult to deduce that the angles of $\Delta A_3 B_3 D_3$ are formed by x, x', y, y' and z, z', so that $\Delta A_3 B_3 C_3$ is similar to ΔABC.

It follows that triangles $A_6 B_6 C_6$, $A_9 B_9 C_9$, $A_{12}, B_{12}, C_{12}, \ldots$ are also similar to $\Delta A_0 B_0 C_0$. (Of course, in special cases $\Delta A_1 B_1 C_1$ itself can be similar to $\Delta A_0 B_0 C_0$.)

Problem 4

S is the centre of b. For each circle c_k denote by O_k the centre and by X_{k+1} the point where it touches c_{k+1}. Let Y_k and Z_k be the projections of O_k and X_k, respectively, onto SZ_1. Denote by r_k the radius of c_k.

r_k can be calculated from the right-angled triangle $SO_k Y_k$ in which the hypotenuse $SO_k = 1 + r_k$, the side $O_k Y_k = 1$ and $SY_k = SZ_k - r_k$.

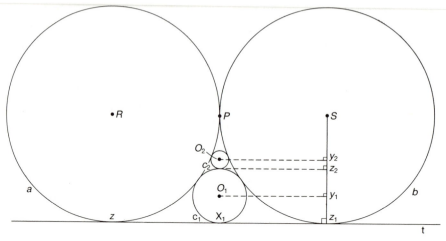

Fig. 1.34

By induction it can be shown that $SZ_k = 1/k$ and that $r_k = 1/[2k(k+1)]$. Thus $d_k = 2r_k = 1/[k(k+1)]$, and we have established the relationship

$$\frac{1}{1\cdot2} + \frac{1}{2\cdot3} + \frac{1}{3\cdot4} + \cdots = 1.$$

Problem 5 (E)

Denote by A, B, C, D, E vertices of the pentagon P_k, by F, G, H, I, J the vertices of the pentagram S_k (which are at the same time the vertices of P_{k+1}) and by K, L, M, N, O the vertices of S_{k+1} as shown in Fig. 1.35.

Each angle of a regular pentagon is $108°$. It is easy to calculate the angles of the isosceles triangles AGB and GBH and deduce that $\angle GAH = \angle AGH$; hence triangle AGH is also isosceles. Thus $GH = AB + BH$, or

$$a_{k+1} = a_k + b_k.$$

Using the isosceles triangle KGI we can express the length of b_{k+1} as

$$b_{k+1} = GK = GI = b_k + a_k + b_k = b_k + a_{k+1}.$$

This proves (a).

To show (b), rewrite the sequence S by expressing each of its elements in terms of a and b. This can be done using (a).

$$S\colon 1a,\ 1b,\ 1a+1b,\ 1a+2b,\ 2a+3b,\ 3a+5b,\ 5a+8b,\ \ldots.$$

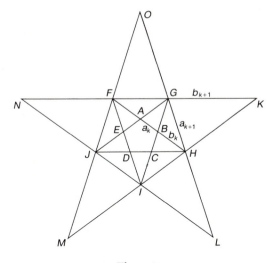

Fig. 1.35

The coefficients of a in each term form the sequence 1, 0, 1, 1, 2, 3, 5, 8, . . . and the coefficients of b form the sequence 0, 1, 1, 2, 3, 5, 8, ... Since each term of S is the sum of the preceding two, it follows that the kth term of S is

$$s_k = f_{k-2}a + f_{k-1}b \qquad \text{for } k = 3, 4, \ldots$$

Remark: The sequence $P_1, S_1, P_2, S_2, \ldots$ can be extended by constructing pentagrams and pentagons *inside* P_1: S_1' is obtained by joining the vertices of P_1; the sides of S_1' determine P_1'; S_2' is obtained by joining the vertices of P_1', and so on. This extends the sequence of the side lengths to

$$\ldots, 5a - 3b, \; -3a + 2b, \; 2a - b, \; -a + b, \; a, \; b, \; a + b, \; a + 2b, \; 2a + 3b, \ldots.$$

Thus

$$s_k' = (-1)^k f_{k+1}a + (-1)^{k-1}f_k b \qquad \text{for } k = 1, 2, \ldots.$$

When k tends to infinity, the length s_k' becomes very small — it tends to 0. This implies that

$$\lim_{k \to \infty} [(-1)^k f_{k+1}a + (-1)^{k-1}f_k b] = 0,$$

so that

$$\lim_{k \to \infty} \frac{f_{k+1}}{f_k} = \frac{b}{a}.$$

Take $a = 1$. Then b can be calculated using the similarity of triangles ABG and AGH (Fig. 1.36):

$$1 : b = b : (1 + b),$$

so that

$$b = \frac{1 + \sqrt{5}}{2}.$$

This leads to the well-known property of the Fibonacci numbers

$$\lim_{k \to \infty} \frac{f_{k+1}}{f_k} = \frac{1 + \sqrt{5}}{2}.$$

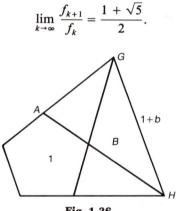

Fig. 1.36

Problem 6 (E)

At the second step the number in the top-left cell will contain the product $a_1^2 \cdot a_5^2 \cdot a_3 \cdot a_7$. Since $(-1)^2 = 1^2 = 1$, this product can be written as $a_3 \cdot a_7$. After

A_2			B_2
$a_3 . a_7$	$a_2 . a_8$	$a_1 . a_9$
$a_4 . a_6$	1	$a_6 . a_4$
$a_1 . a_9$	$a_2 . a_8$	$a_3 . a_7$

(a) D_2 C_2

Fig. 1.37(a)

replacing the square factors in each product in $A_2B_2C_2D_2$ by $+1$, the array in Fig. 1.37(a) is obtained.

The next stage is shown in Fig. 1.37(b).

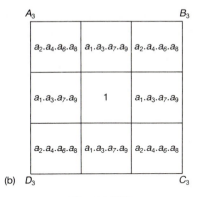

A_3 B_3

$a_2.a_4.a_6.a_8$	$a_1.a_3.a_7.a_9$	$a_2.a_4.a_6.a_8$
$a_1.a_3.a_7.a_9$	1	$a_1.a_3.a_7.a_9$
$a_2.a_4.a_6.a_8$	$a_1.a_3.a_7.a_9$	$a_2.a_4.a_6.a_8$

(b) D_3 C_3

Fig. 1.37(b)

This leads to Fig. 1.37(c).

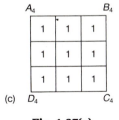

A_4 B_4

1	1	1
1	1	1
1	1	1

(c) D_4 C_4

Fig. 1.37(c)

Thus after at most four steps the initial square $ABCD$ is transformed into a square in which each entry is 1.

Problem 7

If the last digits of two numbers are a and b respectively, then their product will have the same last digit as $a \cdot b$. Thus one finds, step by step, that 7^2 ends with 9, 7^3 with 3, and 7^4 with 1. The last digit of $7^7 = 7^4 \cdot 7^3$ is the same as the last digit of $1 \cdot 3$ which is 3.

Since $(7^7)^2 = 7^7 \cdot 7^7$, its last digit is 9. The last digit in $(7^7)^3$ is the same as the last digit in $9 \cdot 3$, that is 7; the last digit in $(7^7)^4$ is the same as the last digit in $7 \cdot 3$, that is, 1. Hence the last digit in $(7^7)^7$ is the same as the last digit of $7 \cdot 1$ which is 7.

$((7^7)^7)^2 = (7^7)^7 \cdot (7^7)^7$ ends with the last digit of 7×7 which is 9; $((7^7)^7)^3$ ends with 3; $((7^7)^7)^4$ ends with 1, hence $((7^7)^7)^7$ ends with 3.

By continuing in this way, one obtains the following result: If k is an even number, then a_k ends with the digit 3 and if k is odd then a_k ends with 7.

Problem 8

The terms in the sequence can be expressed as

$$10^4 + 1, \ 10^8 + 10^4 + 1, \ldots, \ 10^{4k} + 10^{4(k-1)} + \cdots + 10^4 + 1.$$

Rewrite the general term in the form

$$10^{4k} + 10^{4k-4} + \cdots + 10^4 + 1 = \frac{10^{4(k+1)} - 1}{10^4 - 1}$$

$$= \frac{10^{2(k+1)} + 1}{10^2 + 1} \cdot \frac{10^{2(k+1)} - 1}{10^2 - 1}$$

$$= \frac{(10^{2k+2} + 1) \cdot (10^{2k} + 10^{2k-2} + \cdots + 10^2 + 1)}{101}$$

Since 101 is a prime, it divides at least one of the factors $10^{2k+2} + 1$ and $10^{2k} + \cdots + 10^2 + 1$. If $k > 1$ then both factors are greater than 101. Hence the right-hand side of the above equality is a composite number. Therefore $10^{4k} + 10^{4k-4} + \cdots + 10^4 + 1$ cannot be a prime number if $k > 1$. If $k = 1$ the term reduces to $10^4 + 1$, which is equal to $137 \cdot 73$, hence it is not a prime either.

So, the sequence 10001, 100010001, 1000100010001, . . . contains no prime number.

Problem 9

(a) If A is a one-digit number, then $D(A) = A$. If A has two digits, then the difference

$$A - D(A) = 10a_1 + a_0 - (a_1 + 2a_0) = 9a_1 - a_0$$

is positive except when $a_1 = 1$ and $a_0 = 9$. Thus for any two-digit number A other than 19 the number $D(A)$ is smaller than A.

If A has more than two digits (that is, $n > 1$), then

$$D(A) = a_n + 2a_{n-1} + \cdots 2^n a_0 \le 9 + 2 \cdot 9 + \cdots + 2^n \cdot 9$$

$$= 9(2^{n+1} - 1) < 18 \cdot 2^n < 10^n < A.$$

It follows that the positive numbers A, A_1, A_2, . . . form a decreasing sequence which terminates in a number A_k such that $D(A_k) = A_k$. Then A_k is either 19 or a one-digit number.

(b) $D(A)$ is divisible by 19 whenever A is divisible by 19. To show this form the expression

$$2^n A - D(A) = (20^n - 1)a_n + 2 \cdot (20^{n-1} - 1)a_n + \cdots + 2^{n-1}(20 - 1)a_1.$$

Each summand on the right-hand side of this equality is divisible by 19; thus, if A is divisible by 19, then 19 must divide $D(A)$ as well.

If $A = 19^{85}$, then A, A_1, A_2, . . ., A_k are all divisible by 19. Therefore the number A_k in this sequence for which $D(A_k) = A_k$ is 19.

Problem 10

We shall describe two methods for the solution of this problem (see [90]).

(a) The first method uses vectors.

Place a coordinate system over the chessboard so that the centres of the squares on the board receive integer coordinates. Take the origin as the start for the knight. After the first step the knight can land on any of the eight points with position vectors

$$\pm \begin{pmatrix} p \\ q \end{pmatrix}, \ \pm \begin{pmatrix} p \\ -q \end{pmatrix}, \ \pm \begin{pmatrix} q \\ p \end{pmatrix}, \ \pm \begin{pmatrix} q \\ -p \end{pmatrix}.$$

Thus any point P which is reached by the knight has a position vector of the form

$$\begin{pmatrix} x \\ y \end{pmatrix} = a \begin{pmatrix} p \\ q \end{pmatrix} - b \begin{pmatrix} p \\ q \end{pmatrix} + c \begin{pmatrix} p \\ -q \end{pmatrix} - d \begin{pmatrix} p \\ -q \end{pmatrix}$$
$$+ e \begin{pmatrix} q \\ p \end{pmatrix} - f \begin{pmatrix} q \\ p \end{pmatrix} + g \begin{pmatrix} q \\ -p \end{pmatrix} - h \begin{pmatrix} q \\ -p \end{pmatrix} \tag{1}$$

where a, b, c, d, e, f, g and h are non-negative integers such that

$$s = a + b + c + d + e + f + g + h$$

is the number of steps leading from O to P.

(1) yields the following system of equations:

$$(a - b)p + (c - d)p + (e - f)q + (g - h)q = x$$
$$(a - b)q - (c - d)q + (e - f)p - (g - h)p = y,$$

that is

$$\left.\begin{array}{l} a'p + c'p + e'q + g'q = x \\ a'q - c'q + e'p - g'p = y \end{array}\right\} \qquad (2)$$

where $a' = a - b, c' = c - d, e' = e - f, g' = g - h$. The sum s' of a', c', e' and g' is equal to $s - 2(b+d+f+h)$. Hence s is even if and only if s' is even.

Suppose that after n steps the knight has returned to the origin. In this case $x = y = 0$ and equations (2) can be rewritten as

$$(a' + c')p = -(e' + g')q,$$
$$(a' - c')q = -(e' - g')p.$$

By multiplying the first of these equations by the second and cancelling $p \cdot q$ $(p \neq 0 \neq q)$ we see that

$$(a')^2 - (c')^2 = (e')^2 - (g')^2$$

or

$$(a')^2 - (c')^2 - (e')^2 + (g')^2 = 0. \qquad (3)$$

The numbers $(a')^2 - a'$, $-(c')^2 - c'$, $-(e')^2 - e'$ and $(g')^2 - g'$ are all even. Denote them by $2t$, $2u$, $2v$ and $2w$ respectively, and rewrite (3) as follows:

$$a' + 2t + c' + 2u + e' + 2v + g' + 2w = 0$$

or

$$(a' + c' + e' + g') + 2(t + u + v + w) = 0.$$

The last equation shows that s' is even; hence s is even. It follows that the knight can return to the origin only by making an even number of moves.

(b) The second solution uses a function defined on the set of pairs (x, y), where x and y are integers.

In the coordinate system introduced in the previous solution the centres of the squares have coordinates x, y.

Denote the greatest common divisor of p and q by d and form the numbers p/d and q/d. There are two possibilities.

Case 1. Both p/d and q/d are odd.

In this case define a function $f(x, y)$ for all pairs of integers (x, y) according to the rule

$$f(x, y) = \frac{x}{d}.$$

This function has the property that when the knight makes one move from the square (x, y) to a square (x', y'), then the difference $f(x', y') - f(x, y)$ is an odd number. (In fact, x' is of the form $x \pm p$ or $x \pm q$; so

$$f(x', y') = \frac{x \pm p}{d} = \frac{x}{d} \pm \frac{p}{d} \quad \text{or} \quad f(x', y') = \frac{x \pm q}{d} = \frac{x}{d} \pm \frac{q}{d} \Big).$$

Case 2. One of p/d and q/d is even. (Notice that both numbers cannot be even since d is the greatest common divisor of p and q.)

In this case define f as $f(x, y) = (x + y)/d$. Again it is easy to verify that when the knight makes one move from the square (x, y) to a square (x', y') then the difference $f(x', y') - f(x, y)$ is an odd number.

Let us now return to the knight. Suppose that, starting from $(x_0, y_0) = (0, 0)$, the knight lands first on a square (x_1, y_1), then on (x_2, y_2), and so on, returning after n steps to $(x_n, y_n) = (0, 0)$. After each step the value of f is changed by an odd number. $f(0, 0)$ is even; thus if n were odd, $f(x_n, y_n)$ must have been odd. However, $f(x_n, y_n) = f(0, 0)$; therefore n must be even.

Section 2: Search for patterns

Problem 11

(a) The squares along the boundary can be reached in one direction only (either from the left or from the top), so they will be filled with 1s. Any other square S on the board can be approached from the left and from above; therefore the number in S will be the sum of the numbers in the left and top neighbouring squares.

Thus the numbers on the board can be obtained, step by step, by filling in the squares along the diagonals $d_0, d_1, d_2, \ldots$ marked in Fig. 1.38. The emerging number pattern is the famous Pascal's triangle.

On the other hand, the number in each square can be calculated directly as follows: label the (horizontal) rows and the (vertical) columns of the board by $0, 1, 2, \ldots$. Call a 'step' on the board a move from one square to a neighbouring square. Each path of the rook from A to the square S in the mth row and the nth column consists of $m + n$ steps, n of which must be in the horizontal and m in the vertical direction. It is left to the rook to decide which n of the $m + n$ steps should be horizontal. Therefore:

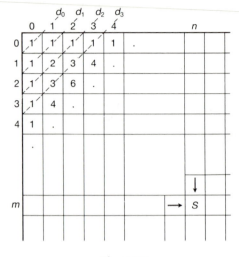

Fig. 1.38

The number of different paths leading from A to S is equal to the number of choices of n out of $m + n$ steps.

There is a well-known formula for the number of choices of s elements out of a set of t elements:

$$C_s^t = \frac{t(t - 1)(t - 2) \cdots (t - s + 2)(t - s + 1)}{s(s - 1) \cdots 3 \cdot 2 \cdot 1},$$

or, more briefly,

$$C_s^t = \frac{t!}{s!(t - s)!},$$

where for any natural number k the symbol $k!$ stands for $k(k - 1)(k - 2) \cdots \cdot 3 \cdot 2 \cdot 1$. By definition $0! = 1$. It follows that the number in the mth row and the nth column of the number pattern is $C_m^{m+n} = (m + n)!/m!n!$.

The numbers C_s^t are also called binomial coefficients; their properties have been extensively investigated (see for example [50]). Here we shall mention only the following interesting relationship between the binomial coefficients and the Fibonacci numbers.

Rewrite Pascal's triangle, as shown in Fig. 1.39. The sums along the dotted diagonals are the Fibonacci numbers 1, 1, 2, 3, 5, 8, 13, ... (This can be proved by induction.)

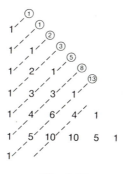

Fig. 1.39

(b) In this case a square S in the interior of the board can be reached from three squares, as shown in Fig. 1.40. So, again, the number pattern can be constructed step by step, proceeding along the diagonals $d_0, d_1, d_2, \ldots$.

Our next task is to find an expression for the number of paths leading from A to the square S in the mth row and the nth column.

Suppose that the king takes r steps diagonally downwards; then the number of horizontal steps taken must be $n - r$, and the number of vertical steps $m - r$. The r diagonal, $n - r$ horizontal and $m - r$ vertical steps can be taken in any order. Therefore, the number of those paths that involve r

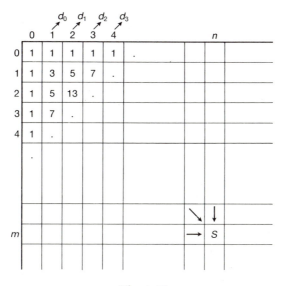

Fig. 1.40

diagonal steps is equal to the number of permutations of the total of $r + (n - r) + (m - r) = n + m - r$ steps, of which r are of one kind, $n - r$ of a second kind and $m - r$ of a third kind. It is well known that this number is equal to

$$\frac{(n + m - r)!}{r!(n - r)!(m - r)!}.$$

But the king can take any number of diagonal steps ranging from 0 to the smaller one of the numbers m and n. Thus the total number of paths leading from A to S is given by the sum

$$\sum_{r=0}^{\min(m, n)} \frac{(n + m - r)!}{r!(n - r)!(m - r)!}$$

By rewriting the number pattern as shown in Fig. 1.41 and by adding the numbers along the dotted diagonals we obtain the number sequence 1, 1, 2, 4, 7, 13, 24, ...

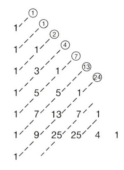

Fig. 1.41

These numbers are nicknamed *tri*bonacci numbers because each of them, beginning with the fourth, is the sum of the preceding three.

(c) Finally, the number pattern in part (c) of Problem 11 is shown in Fig. 1.42.

The rule of forming the individual terms in the pattern is not as straightforward as in the two previous cases. A connection with the trinomial coefficients will be pointed out on p. 76.

The link with the tribonacci numbers can be established in the same way as in part (b).

				1				
			1	1	1			
		1	2	3	2	1		
	1	3	6	7	6	3	1	
1	4	10	16	19	16	10	4	1
.	.	.	.	.	.	.	.	.

Fig. 1.42

Problem 12

(a) Consider the nth square. The numbers in the first row add up to

$$r_1 = 1 + 2 + 3 + \cdots + n = \frac{n(n + 1)}{2},$$

which is the nth triangular number; the numbers in the ith row add up to

$$r_i = i \cdot r_1 \qquad \text{for } i = 1, 2, \ldots, n.$$

Thus the sum of all numbers in the nth square is

$$S_n = \frac{n(n + 1)}{2} + 2\frac{n(n + 1)}{2} + \cdots + n\frac{n(n + 1)}{2} =$$

$$= \frac{n(n + 1)}{2}(1 + 2 + \cdots + n) = \left(\frac{n(n + 1)}{2}\right)^2.$$

In the nth corridor the numbers add up to

$$c_n = 2[n + 2n + \cdots + (n-1)n] + n^2 = n^3.$$

Since $c_1 + c_2 + \cdots + c_n = s_n$, these results show that the sum of the first n cube numbers is equal to the square of the nth triangular number.

(b) Denote the number in the mth row and in the nth column of the array by a_{mn}.

In the mth row the numbers form an arithmetic progression with first term 1 and common difference $m + 1$, so that

$$a_{mn} = 1 + (n - 1)(m + 1). \tag{4}$$

It follows that:

(i) $\qquad\qquad\qquad a_{nn} = 1 + (n - 1)(n + 1) = n^2;$

that is,

All numbers in the array along the diagonal containing a_{11} and a_{22} are square numbers.

(ii) All numbers in the square $S_{k\ell}$ (Fig. 1.43) add up to a square number.

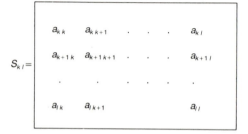

$$S_{k\,l} = \begin{array}{cccc} a_{k\,k} & a_{k\,k+1} & \cdot \quad\cdot\quad\cdot & a_{k\,l} \\[1mm] a_{k+1\,k} & a_{k+1\,k+1} & \cdot \qquad\cdot & a_{k+1\,l} \\[2mm] \cdot & \cdot \quad\cdot\quad\cdot\quad\cdot & & \cdot \\[1mm] a_{l\,k} & a_{l\,k+1} & & a_{l\,l} \end{array}$$

Fig. 1.43

Indeed, the sum of the numbers in the first row of $S_{k\ell}$ is

$$r_1 = \frac{\ell - k + 1}{2}(a_{kk} + a_{k\ell});$$

in the second row,

$$r_2 = \frac{\ell - k + 1}{2}(a_{k+1\,k} + a_{k+1\,\ell});$$

$$\vdots$$

in the ℓth row

$$r_\ell = \frac{\ell - k + 1}{2}(a_{\ell k} + a_{\ell\ell}).$$

Thus the sum of all numbers in $S_{k\ell}$ is:

$$s = \frac{\ell - k + 1}{2}(a_{kk} + a_{k+1\,k} + \cdots + a_{\ell k})$$

$$+ \frac{\ell - k + 1}{2}(a_{k\ell} + a_{k+1\,\ell} + \cdots + a_{\ell\ell}). \qquad (5)$$

The numbers a_{kk}, $a_{k+1\,k}$, . . ., $a_{\ell k}$ form an arithmetic progression; therefore their sum is equal to

$$\frac{\ell - k + 1}{2} (a_{kk} + a_{\ell k}),$$

or, in view of (4), to

$$\frac{\ell - k + 1}{2} (k^2 + \ell k + \ell - k). \tag{6}$$

The numbers $a_{k\ell}$, $a_{k+1\,\ell}$, . . ., $a_{\ell\ell}$ also form an arithmetic progression; their sum is equal to

$$\frac{\ell - k + 1}{2} (a_{k\ell} + a_{\ell\ell}) = \frac{\ell - k + 1}{2} (\ell^2 + \ell k + k - \ell). \tag{7}$$

By inserting the expressions (6) and (7) into (5) one finds

$$S = \frac{\ell - k + 1}{2} \cdot \frac{\ell - k + 1}{2} (k^2 + 2\ell k + \ell^2)$$

$$= \left(\frac{\ell - k + 1}{2}\right)^2 (k + \ell)^2 = \left[\frac{(\ell - k + 1)(\ell + k)}{2}\right]^2,$$

that is, a square number.

(c) We shall show that $2k + 1$ is *not* a prime number if and only if k is in the sieve.
(i) Suppose that k is in the sieve, say in the mth row and nth column. Then k is equal to

$$k = (1 + 3m) + (n - 1)(2m + 1) = (2n + 1)m + n. \tag{8}$$

Thus

$$2k + 1 = 2[(2n + 1)m + n] + 1 = 2(2n + 1)m + 2n + 1$$

$$= (2n +)(2m + 1).$$

This number is not a prime, since both $2n + 1$ and $2m + 1$ are greater than 1.
(ii) Suppose that $2k + 1$ is not a prime. Then $2k + 1$ is the product of two odd numbers $2a + 1$ and $2b + 1$:

$$2k + 1 = (2a + 1)(2b + 1) = 2(2ab + a + b) + 1,$$

and

$$k = 2ab + a + b = (2b + 1)a + b.$$

But then k is of the form (8), and therefore appears in the table.

(d) In the kth corridor of the array the numbers add up to

$$S_k = 1 + 2 + \cdots + (k - 1) + k \cdot k$$

$$= \frac{(k - 1)k}{2} + k^2 = \tfrac{3}{2}k^2 - \frac{k}{2};$$

hence the sum of all numbers in the array is

$$S = S_1 + S_2 + \cdots + S_n$$

$$= \tfrac{3}{2}(1^2 + 2^2 + \cdots + n^2) - \tfrac{1}{2}(1 + 2 + \cdots + n)$$

$$= \tfrac{3}{2}S_{2,n} - \tfrac{1}{2}\frac{n(n + 1)}{2}. \tag{9}$$

On the other hand, in each of the n rows of the array the numbers add up to

$$S_{1,n} = 1 + 2 + \cdots + n = \frac{n(n + 1)}{2};$$

therefore

$$S = n \cdot \frac{n(n + 1)}{2}. \tag{10}$$

Combining (9) and (10), it follows that

$$\tfrac{3}{2}S_{2,n} - \tfrac{1}{4}n(n + 1) = n \cdot \frac{n(n + 1)}{2}$$

or

$$S_{2,n} = \frac{n(n + 1)(2n + 1)}{6}.$$

Remark: A similar array can be used to find the sum of the cubes

$$S_{3,n} = 1^3 + 2^3 + \cdots + n^3$$

if $S_{2,n}$ and $S_{1,n}$ are known:

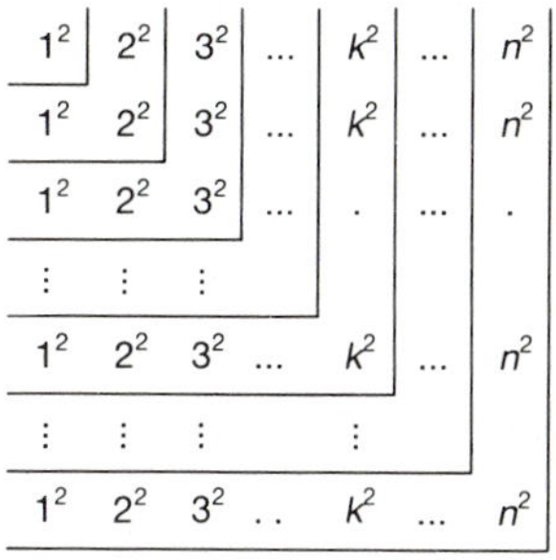

Fig. 1.44

It is left to the reader to verify that

$$S_{3,n} = \frac{n^2(n + 1)^2}{4}.$$

Problem 13 (E)

The numbers $a_1 + b_1, a_1 + b_2, \ldots, a_{10} + b_{10}$ can be divided into the following ten disjoint subsets S_i, where $i = 1, 2, \ldots, 10$:

$$S_i = \{a_1 + b_i, a_2 + b_{i+1}, a_3 + b_{i+2}, \ldots, a_{10}b_{i+9}\}$$

where the sums $i + 1, i + 2, \ldots, i + 9$ are taken modulo 10.

In each S_i the sum of the elements is

$$a_1 + a_2 + a_3 + \cdots + a_{10} + b_1 + b_2 + b_3 + \cdots + b_{10}.$$

Problem 14 (E)

Let ABC be an isosceles triangle with equal angles at B and C. Suppose that $\triangle ABC$ is divided by a straight line ℓ into two isosceles triangles. We distinguish

(a) ℓ divides one of the equal angles, say the angle at B;

(b) ℓ divides the angle at the apex A.

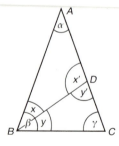

Fig. 1.45

Case (a)   Denote the angles of $\triangle ABC$ at A, B, C, by α, β and γ respectively. Let D be the point where ℓ meets AC, and put

$$\angle ABD = x, \qquad \angle ADB = x'$$

$$\angle DBC = y, \qquad \angle BDC = y'.$$

Since the isosceles triangles ABD and DBC might have their equal angles in various positions, the following nine subcases must be looked at:

(1) $\alpha = x$ and $y = \gamma$, (4) $\alpha = x'$ and $y = \gamma$, (7) $x = x'$ and $y = \gamma$,
(2) $\alpha = x$ and $y = y'$, (5) $\alpha = x'$ and $y = y'$, (8) $x = x'$ and $y = y'$,
(3) $\alpha = x$ and $y' = \gamma$, (6) $\alpha = x'$ and $y' = \gamma$, (9) $x = x'$ and $y' = \gamma$.

Seven of the above possibilities can be ruled out:

(7), (8) and (9) cannot occur since $x' > \gamma = \beta > x$;
(1) and (4) cannot occur either because $y < \beta = \gamma$;
(6) also leads to a contradiction — namely

$$x' + y' = 180°; \text{ however, } \alpha + \gamma < 180°.$$

(5) cannot hold since it would imply that $x' + y' < 180°$.

In the remaining two cases, using the fact that in any triangle all angles add up to $180°$, the angles of $\triangle ABC$ can be easily calculated:

In case (2) $\alpha = \dfrac{180°}{7}$, $\beta = \gamma = \dfrac{540°}{7}$;

in case (3) $\alpha = 36°$, $\beta = \gamma = 72°$.

Case (b)   The line ℓ cuts BC at a point E. Denote the angles of $\triangle ABC$ at A, B, C by α, β and γ respectively, and put $u = \angle BAE$, $v = \angle EAC$, $u' = \angle BEA$ and $v' = \angle AEC$.

Again, there are 9 possibilities. It is easy to verify that only the following cases can occur:

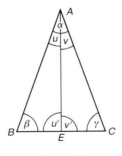

Fig. 1.46

(1) $u = \beta$ and $v = \gamma$; this implies that $a = 90°$, $\beta = \gamma = 45°$.
(2) $u = \beta$ and $v = v'$; here $\alpha = 108°$, $\beta = \gamma = 36°$.
(3) $v = \gamma$ and $u = u'$; this case is symmetrical to case (2); again $\alpha = 108°$, $\beta = \gamma = 36°$.

Problem 15 (E)

The answer to the question is 'Yes'. The division of a regular polygon with 6, 8, 10, . . . sides into rhombuses can be neatly demonstrated by using linkages as shown in Fig. 1.47.

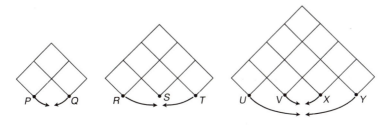

Fig. 1.47

The linkages in Fig. 1.47 are made from rods of equal length, forming 3, 6 and 10 squares respectively. If the vertices P and Q are pulled together into a single point, then the perimeter of the shape can be adjusted so that a regular hexagon is formed. Similarly by pulling together the points R, S, T and adjusting the perimeter, a regular octagon is formed, and by pulling together U, V, X, Y a regular decagon is obtained. Each of these shapes is then divided into rhombuses; their numbers are 3, 6 and 10 — that is, triangular numbers.

In general, from a linkage constructed in a similar manner from equal rods forming $(n - 1)n/2$ squares, a regular polygon with $2n$ sides can be constructed. This polygon is then automatically divided into $(n - 1)n/2$ rhombuses.

Problem 16 (E)

First solution.   Denote the squares with side lengths 1, 1, 2, 3, . . . by S_1, S_2, S_3, S_4, . . . respectively.

Figure 1.48 shows the diagonals DA, GC, FH and KJ of the squares S_1, S_3, S_5 and S_7. They are all parallel; consequently, GC, FH and KJ can be extended to construct a sequence of parallelograms: $P_1 = BADC$, $P_3 = BFEG$, $P_5 = EJIH$ and $P_7 = IKTL$. In this sequence any two consecutive parallelograms are in perspective: P_1 and P_3 from centre B, P_3 and P_5 from centre E, P_5 and P_7 from centre I. The centres of S_2, S_4, S_6, S_8 are the midpoints of the diagonals of P_1, P_3, P_5 and P_7 respectively; it is easy to deduce that they are all on a common line ℓ_1.

By induction it follows that ℓ_1 contains the centres of all squares in the sequence S_2, S_4, S_6, S_8, S_{10}, ...

Similarly, since AC, FG, HJ, and KL are all parallel, they can be produced to construct the parallelograms P_2, P_4, P_6, and P_8 (Fig. 1.49). The parallelograms P_2 and P_4 are in perspective from centre M, P_4 and P_6 are in perspective from centre P, and P_6, and P_8 are in perspective from R. The centres of the squares S_1, S_3, S_5, and S_7 are the midpoints of the diagonals of P_2, P_4, P_6 and P_8, and so lie on a common line ℓ_2.

By induction it follows that ℓ_2 contains the centre of each square in the sequence S_1, S_3, S_5, ...

Since the parallelograms P_1 and P_2 are congruent and their corresponding sides meet at right angles, their corresponding diagonals must also be perpendicular to one another. Thus ℓ_1 and ℓ_2 are perpendicular to each other.

Second solution.   Denote the centre of S_i by O_i for $i = 1, 2, 3, ...$ Draw a coordinate system with origin O_1 and axes parallel to the sides of the squares as shown in Fig. 1.50.

Denote the coordinates of O_i and x_i and y_i. It is left to the reader to verify the relationships

$$x_{4n+7} - x_{4n+3} = \tfrac{1}{2}(f_{4n+7} + f_{4n+3})$$

and

$$y_{4n+7} - y_{4n+3} = \tfrac{1}{2}(f_{4n+7} - 2f_{4n+4} - f_{4n+3}),$$

where f_i is the ith Fibonacci number.

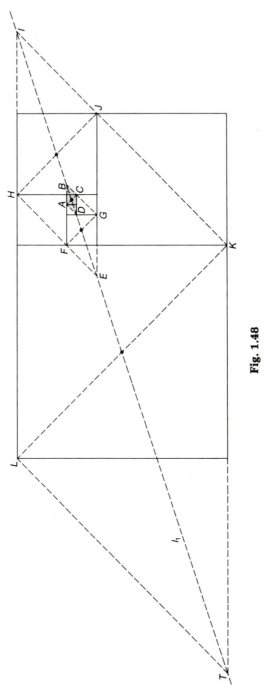

Fig. 1.48

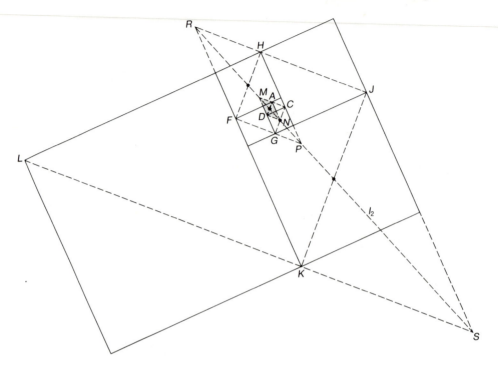

Fig. 1.49

The slope of the line through O_{4n+7} and O_{4n+3} is

$$g = \frac{y_{4n+7} - y_{4n+3}}{x_{4n+7} - x_{4n+3}} = \frac{f_{4n+7} - 2f_{4n+4} - f_{4n+3}}{f_{4n+7} + f_{4n+3}}.$$

Since $f_{n+2} = f_{n+1} + f_n$ for $n = 1, 2, \ldots$, it follows that $f_{4n+7} = 3f_{4n+4} + 2f_{4n+3}$ and $g = \frac{1}{3}$.

Thus the centres O_3, O_7, O_{11}, O_{15}, ... lie on the line ℓ_2 with equation $y = \frac{1}{3}x$.

By similar arguments one can show that:

(i) O_1, O_5, O_9, ... belong to ℓ_2,

(ii) O_2, O_4, O_6, ... are on the line ℓ_1 with equation $y = -3x + 1$.

Since the slopes of ℓ_2 and ℓ_1 are $\frac{1}{3}$ and -3, the lines meet at right angles.

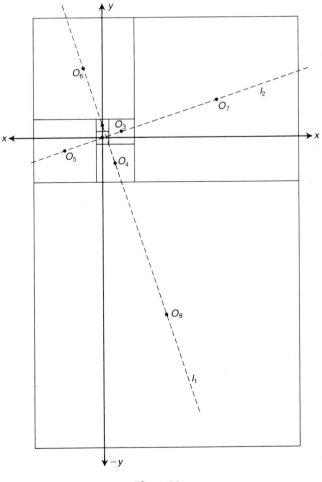

Fig. 1.50

Problem 17

(a) x, y, z and w must be positive integers adding up to 12. Represent the 12 units by 12 dots in a row:

Fig. 1.51

Suppose that $x = 3$, $y = 1$, $z = 6$ and $w = 2$. x can be represented by the first three dots in the row, y by the next dot, z by the following 6 dots and w by the final 2. These four groups of dots can be separated by inverting division lines into the appropriate gaps between the dots (Fig. 1.52).

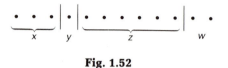

Fig. 1.52

Any other set of solutions of the equation $x + y + z + w = 12$ can be represented in a similar way.

There are 11 gaps between the twelve dots. By inserting a division line in any three of these gaps the dots will be divided into four non-empty sets; the number of dots in the first set will be equal to x, in the second set to y, in the third to z, and in the fourth to w.

Hence the number of all positive integer solutions of the equation $x + y + z + w = 12$ is equal to the number of ways in which 3 gaps can be chosen out of 11, to be filled with division lines. This number is known to be

$$C_3^{11} = \frac{11 \cdot 10 \cdot 9}{3 \cdot 2 \cdot 1} = 165.$$

(b) Similarly, the number of positive integer solutions of the equation

$$x_1 + x_2 + \cdots + x_n = k$$

is equal to

$$C_{n-1}^{k-1} = \frac{(k-1)!}{(n-1)!(k-n)!}$$

where $a! = a(a-1)(a-2) \cdots 3 \cdot 2 \cdot 1$.

Clearly, if n is greater than k, then the equation has no solution in positive integers.

Problem 18

Rewrite $2^{2^n} + 1$ in the form $(2^{2^n} - 1) + 2$. The difference $2^{2^n} - 1$ can be factored:

$$2^{2^n} - 1 = (2^{2^{n-1}} + 1)(2^{2^{n-1}} - 1).$$

Moreover,

$$2^{2^{n-1}} - 1 = (2^{2^{n-2}} + 1)(2^{2^{n-2}} - 1)$$
$$= (2^{2^{n-2}} + 1)(2^{2^{n-3}} + 1) \ldots (2^{2^m} + 1) \ldots (2^2 + 1)(2 + 1)(2 - 1).$$

Hence

$$2^{2^n} + 1 = \underbrace{(2^{2^{n-1}} + 1)(2^{2^{n-2}} + 1) \cdots (2^{2^m} + 1) \cdots (2^2 + 1)(2 + 1)}_{A} + 2. \quad (11)$$

Suppose that $2^{2^n} + 1$ and $2^{2^m} + 1$ have a common factor $d > 1$. Take $n > m$. Then d divides the left-hand side of the equality (11) and also the term A on the right-hand side. Therefore d must divide 2. Since $d > 1$, this implies that d must be 2. However, $2^{2^n} + 1$ is an odd number, and cannot be divisible by 2.

This shows that $2^{2^n} + 1$ and $2^{2^m} + 1$ are relatively prime for any positive integers m and n, $m \neq n$.

Problem 19

The first step is to show that a tree diagram consisting of all fractions a/b between 0 and 1 (a and b are relatively prime, and $a < b$) must start with the fraction $\frac{1}{2}$.

Indeed, suppose that $\frac{1}{2}$ is the successor of a fraction p/q; then

$$\frac{1}{2} = \frac{p}{p + q} \quad \text{or} \quad \frac{1}{2} = \frac{q}{p + q},$$

that is $p = q = 1$, and $p/q = 1$. But 1 cannot be contained in the diagram.

The tree-diagram T starting with $\frac{1}{2}$ is shown in Fig. 1.53.

The next step is to verify that all elements of T are fractions in their simplest terms, less than 1.

Clearly, $a/(a + b)$ and $b/(a + b)$ are both less than 1. Therefore all immediate and remote successors of $\frac{1}{2}$ are fractions less than 1. Moreover, the successors of a reduced fraction a/b are also fractions in their simplest terms; otherwise the common divisor $d > 1$ of a and $a + b$, or of b and $a + b$, would divide both a and b – leading to a contradiction.

Finally, one has to show that any fraction $p/q < 1$, in its simplest terms, is contained in T.

$\frac{1}{2}$ is in T, by construction.

For $p/q \neq \frac{1}{2}$ there are two cases: (a) $p/q < \frac{1}{2}$ and (b) $p/q > \frac{1}{2}$. In case (a) $q - p > p$, form the fraction $P(p/q) = p/(q - p)$. In case (b) $p > q - p$, define the fraction $P(p/q) = (q - p)/p$.

For any fraction $p/q < 1$ in its simplest term construct the sequence

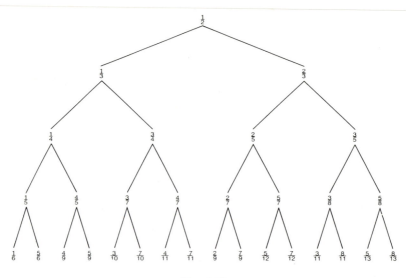

Fig. 1.53

$$\frac{p}{q},\, P\!\left(\frac{p}{q}\right),\, P^2\!\left(\frac{p}{q}\right) = P\!\left(P\!\left(\frac{p}{q}\right)\right),\, P^3\!\left(\frac{p}{q}\right) = P\!\left(P^2\!\left(\frac{p}{q}\right)\right),\,\ldots\ldots(12)$$

The denominator of $P(p/q)$ is smaller than the denominator of p/q (in case (a) $q - p < q$, and in case (b) $p < q$). Thus the denominators of the fractions in (12) form a decreasing sequence. This sequence must end with 2. Hence sequence (12) terminates with the fraction $\frac{1}{2}$.

Suppose that $P^k(p/q) = \frac{1}{2}$. Then it is easy to see that $P^{k-1}(p/q)$ is a successor of $\frac{1}{2}$ in T. Similarly $P^{k-2}(p/q)$ is a successor of $P^{k-1}(p/q)$ in T, $P^{k-3}(p/q)$ a successor of $P^{k-2}(p/q)$, and so on. In other words, there is a chain of successors leading from $\frac{1}{2}$ to p/q, hence p/q is in T.

Problem 20

Any natural number n can be expressed in the form $m = 2^i t$, where t is the greatest odd divisor of m and i is a non-negative integer.

Suppose that the players called out the greatest odd divisors of the numbers in the set $S = \{1, 2, 3, \ldots, n\}$ and then stopped the game.

S can be partitioned into the following disjoint subsets:

S_0 consisting of the odd numbers 1, 3, 5, . . . not greater than n,
S_1 consisting of the numbers 2·1, 2·3, 2·5, . . . not greater than n,

S_2 consisting of the numbers $2^2 \cdot 1$, $2^2 \cdot 3$, $2^2 \cdot 5$, . . . not greater than n,

⋮

S_k consisting of the numbers $2^k \cdot 1$, $2^k \cdot 3$, $2^k \cdot 5$, . . . not greater than n,

where k is a fixed integer, $k \geq 0$.

It is easy to analyse the outcome of the game for each of the subsets S_i (Fig. 1.54).

$m \in S_i$	$2^i \cdot 1$	$2^i \cdot 3$	$2^i \cdot 5$	$2^i \cdot 7$	$2^i \cdot 9$	$2^i \cdot 11$	...
Peter's gain	0	1	0	1	0	1	...
Paul's gain	1	0	1	0	1	0	...

Fig. 1.54

When calling out the divisors t for numbers m in S_i, the players alternate in receiving £1. Since Paul is the first to receive £1 by calling out $2^i \cdot 1$, it follows that:

1. In the part of the game corresponding to S_i for all $i = 0, 1, 2, . . ., k$ Peter can gain no more pounds than Paul.

    Moreover, there is at least one number $j \leq k$ such that $2^j \leq n < 3 \cdot 2^j$ Therefore:

2. In the part of the game corresponding to S_j Paul receives £1 and Peter does not receive anything.

    (1) and (2) imply that: Regardless of when the game stops, Paul will always have at least £1 more than Peter.

Section 3: Exceptions and special cases

Problem 21 (E)

Put $a = CB$, $b = AC$, $x = MB'$ and $y = MA'$ (Fig. 1.55). Since the triangles $AB'M$ and ACB are similar, their corresponding sides are proportional:

$$a : b = x : (b - y). \tag{13a}$$

The perimeter of the rectangle $MA'CB'$ is given by

$$2x + 2y = 2s. \tag{13b}$$

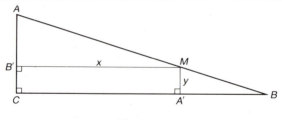

Fig. 1.55

From 13a and 13b it follows that

$$x = \frac{a(s - b)}{a - b} \quad \text{and} \quad y = \frac{b(a - s)}{a - b}$$

The expressions for x and y show that $a - b$ cannot be 0, hence there is no solution to the problem if $a = b$.

Suppose that $a > b$; since x and y must be positive numbers, the inequalities $a > s$ and $s > b$ must be satisfied. Under these conditions the distance of the point M from the vertex A is given by the formula

$$AM = \frac{(s - b)AB}{a - b} = \frac{(s - b)\sqrt{(a^2 + b^2)}}{a - b}.$$

Thus the problem has a unique solution if $a > b$ and $a > s > b$. Similarly, the problem has a unique solution if $b > a$ and $b > s > a$. Otherwise there is no solution.

Problem 22 (E)

(a) Denote the angles of a convex n-gon by $\alpha_1, \alpha_2, \ldots, \alpha_n$. Suppose that

$$\alpha_1 \geq \alpha_2 + \alpha_3 + \cdots + \alpha_n. \tag{14}$$

Bear in mind that in a convex n-gon all angles add up to $(n - 2)180°$. By adding α_1 to both sides of (14) we see that

$$2\alpha_1 \geq \alpha_1 + \alpha_2 + \cdots + \alpha_n = (n - 2)180°.$$

Hence

$$\alpha_1 \geq (n - 2)90°.$$

Since the polygon is convex, $\alpha_1 < 180°$, so that

$$(n - 2)90° < 180°,$$

that is

$$n - 2 < 2$$

or

$$n < 4.$$

Thus, the only convex polygon in which one angle can be equal to or greater than the sum of the remaining angles is the triangle.

Fig. 1.56

(b) In the polygon denote the external angle complementing the internal angle α_i by β_i.

Since $\beta_i = 180° - \alpha_i$ for each $i = 1, \ldots, n$, the sum of all external angles is

$$E_n = \beta_1 + \beta_2 + \cdots + \beta_n = (180 - \alpha_1) + (180 - \alpha_2) + \cdots + (180 - \alpha_n)$$
$$= n \cdot 180° - (\alpha_1 + \alpha_2 + \cdots + \alpha_n)$$
$$= n \cdot 180° - (n - 2)180° = 360°.$$

Let k be the number of acute angles in the n-gon. Then the polygon has k obtuse external angles whose sum is greater than $k \cdot 90°$. Since $(90k)°$ cannot exceed $360°$, this implies that k cannot be greater than 3. The only convex n-gon with n acute angles is the acute-angled triangle.

Problem 23 (E)

(a) In a parallelogram adjacent angles add up to $180°$. Therefore the triangles *AKD*, *ALB*, *CMB* and *DNC* formed by the bisectors of the

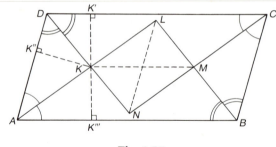

Fig. 1.57

parallelogram *ABCD* are right-angled (Fig. 1.57). This implies that the quadrilateral *KLMN* is a rectangle.

(b) A rectangle is a square exactly when its diagonals meet at right angles. The endpoint *K* of the diagonal *KM* is on the bisector of the angle ∢*CDA*; therefore, the distances *KK'* and *KK"* of *K* from the sides *DC* and *DA* are equal. At the same time *K* is on the bisector of ∢*DAB*; thus the distances *KK"* and *KK'''* of *K* from *DA* and *AB* are equal.

This implies that the distances of *K* from the parallel sides *DC* and *AB* are the same.

Using the same arguments one can prove that the endpoint *M* of *KM* is at equal distances from *DC* and *AB*.

Thus *KM* is parallel to *AB*. Similarly, the diagonal *LN* is parallel to *AD*. It follows that *KM* is perpendicular to *NL* if and only if *AB* is perpendicular to *AD*.

In other words *KLMN* is a square if and only if *ABCD* is a rectangle.

Problem 24

Suppose that the mouseholes form a triangle *ABC* (Fig. 1.58). All points in the plane, at equal distances from *A* and *B*, are on the perpendicular bisector

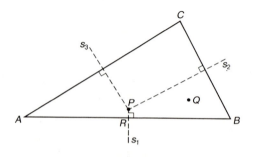

Fig. 1.58

s_1 of AB. Similarly, all points in the plane, at equal distances from B and C, are on the perpendicular bisector s_2 of BC, and at equal distances from C and A on the bisector s_3 of AC. All the bisectors meet at a common point P.

Take a point Q in the plane, different from P. If Q is in the region between s_1 and s_2, then the distances AQ and CQ are greater than BQ. Suppose that $AQ > CQ$. Then A is the furthest mousehole from the cat sitting at Q. This distance can be minimized if the cat moves to a point R on s_1. If R is different from P then CR is the greatest of the three distances CR, AR, BR; this distance can be minimized by moving over to the point which is at equal distances from all the vertices of $\triangle ABC$, that is, to point P.

The problem arises when the triangle ABC has an obtuse angle. This implies that the perpendicular bisectors meet outside the triangle and the regions between the bisectors overlap (Fig. 1.59).

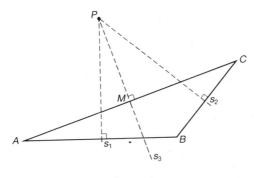

Fig. 1.59

In this case the cat should sit at the midpoint M of the longest side, say AC. In this position $MB < MC$ and $MC = MA$. For any other point N in the plane at least one of the distances AN and CN would be greater than AM.

When $\triangle ABC$ is right-angled, then the bisectors of the sides meet at the midpoint M of the hypotenuse. Hence in this case M provides the solution to the problem.

Finally, the mouseholes could lie on a common straight line with, say, B between A and C. In this case the cat should sit at the midpoint of AC.

Problem 25

The equation

$$\{\sqrt{[a + \sqrt{(a^2 - 1)}]}\}^x + \{\sqrt{[a - \sqrt{(a^2 - 1)}]}\}^x = 2a \tag{15}$$

has real roots only if the expressions $a^2 - 1$, $a - \sqrt{(a^2 - 1)}$ and $a + \sqrt{(a^2 - 1)}$ are all non-negative, that is, if $a \geq 1$.

Therefore we must solve (15) under the assumption that $a \geq 1$.
Notice that

$$\sqrt{[a + \sqrt{(a^2 - 1)}]} \cdot \sqrt{[a - \sqrt{(a^2 - 1)}]} = \sqrt{[a^2 - (a^2 - 1)]} = 1;$$

hence if

$$\{\sqrt{[a + \sqrt{(a^2 - 1)}]}\}^x = y \tag{16}$$

then equation (15) reduces to

$$y + \frac{1}{y} = 2a. \tag{17}$$

The solutions of (17) are

$$y_{1,2} = a \pm \sqrt{(a^2 - 1)}.$$

Substituting these in (16) and taking logarithms, one finds that

$$x_{1,2} = \frac{\log y_{1,2}}{\log\sqrt{[a + \sqrt{(a^2 - 1)}]}}. \tag{18}$$

The expressions (18) are defined only if $\log\sqrt{[a + \sqrt{(a^2 - 1)}]} \neq 0$, that is, if $a + \sqrt{(a^2 - 1)} \neq 1$, in other words if $a \neq 1$.
If $a \neq 1$, then

$$x_1 = \frac{\log[a + \sqrt{(a^2 - 1)}]}{\log\sqrt{[a + \sqrt{(a^2 - 1)}]}} = 2$$

and

$$x_2 = \frac{\log[a - \sqrt{(a^2 - 1)}]}{\log\sqrt{[a + \sqrt{(a^2 - 1)}]}} = \frac{\log \dfrac{1}{a + \sqrt{(a^2 - 1)}}}{\log \sqrt{[a + \sqrt{(a^2 - 1)}]}} = -2.$$

For both values $x = \pm 2$ the left-hand side of the original equation (15) is equal to $2a$.
Thus:

1. For $a > 1$ equation (15) has two solutions: $x_1 = 2$ and $x_2 = -2$.
2. If $a = 1$, then (15) reduces to $1^x + 1^x = 2$, which is true for any real number x.

Problem 26

Any quadrilateral $ABCD$ can be divided by one of its diagonals into two triangles. Let this diagonal be AC. Then the area $\mathbb{A}$ of $ABCD$ is

$$\mathbb{A} = \tfrac{1}{2} ab \sin \beta + \tfrac{1}{2} cd \sin \delta, \qquad (19)$$

where $\beta < 180°$ and $\delta < 180°$.

AC can be calculated in two ways by applying the cosine rule to the triangles ABC and ACD:

$$AC^2 = a^2 + b^2 - 2ab \cos \beta$$

and

$$AC^2 = c^2 + d^2 - 2 \cdot cd \cos \delta.$$

Hence

$$a^2 + b^2 - c^2 - d^2 = 2ab \cos \beta - 2dc \cos \delta. \qquad (20)$$

Combining (19) and (20) we find that

$$(4\mathbb{A})^2 + (a^2 + b^2 - c^2 - d^2)^2 = (2ab \sin \beta + 2cd \sin \delta)^2$$
$$+ (2ab \cos \beta - 2cd \cos \delta)^2. \quad (21)$$

Since $\sin^2 x + \cos^2 x = 1$ for any angle x, the right-hand side of (21) can be simplified:

$$16\mathbb{A}^2 + (a^2 + b^2 - c^2 - d^2)^2 = 4a^2b^2 + 4c^2d^2$$
$$- 8abcd(\cos \beta \cos \delta - \sin \beta \sin \delta),$$

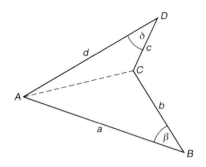

Fig. 1.60

that is

$$16A^2 = 4a^2b^2 + 4c^2d^2 - (a^2 + b^2 - c^2 - d^2)^2 - 8abcd \cos(\beta + \delta). \quad (22)$$

a, b, c, d are given: therefore $16A^2$ will be maximal when $\cos(\beta + \delta)$ is minimal. This is the case when $\cos(\beta + \delta) = -1$, that is, when $\beta + \delta = 180°$.

The quadrilaterals in which the opposite angles add up to $180°$ are the cyclic quadrilaterals. Thus among all quadrilaterals $ABCD$ with given side lengths AB, BC, CD, DA the cyclic quadrilateral has the largest area.

Problem 27

(a) Note that if there is at least one square among the numbers $a^m + a^k$ for some given a, then there are infinitely many square numbers of this form. In fact, if $a^m + a^k = b^2$, then for any $n = 1, 2, \ldots$ the number

$$a^{m+2n} + a^{k+2n} = a^{2n}(a^m + a^k) = a^{2n}b^2 = (a^n b)^2$$

is also a square.

If $a = 2$ then $2^5 + 2^2 = 6^2$; thus there are infinitely many squares of the form $2^m + 2^k$.

(b) If $a = 3$ then $3^3 + 3^2 = 6^2$; hence there are infinitely many squares of the form $3^m + 3^k$.

(c) $a = 4$:

Suppose that $4^m + 4^k = b^2$ for some m and k. Let $m > k$. Then $4^{m-k} + 1 = (b/2^k)^2$, leading to the equation

$$\left(\frac{b}{2^k} - 2^{m-k}\right)\left(\frac{b}{2^k} + 2^{m-k}\right) = 1,$$

which cannot hold.

Hence there is no square among the numbers $4^m + 4^k$.

$a = 5$:

Let $m > k > 0$. 5^m and 5^k for all $m > k > 1$ end with the digits 25. Therefore $5^m + 5^k$ ends with 30 if $k = 1$ and with 50 otherwise. This means that $5^m + 5^k$ cannot be a square number, since in any square number with last digit 0 the digit before the last is also 0.

$a = 6$; $m > k > 0$:

Any power of 6 greater than 0 ends with the digit 6. Therefore $6^k + 6^m$ ends with the digit 2. However, the last digits of all square number are 0, 1, 4, 5, 6 or 9. Thus there is no square number of the form $6^k + 6^m$.

$a = 7$:

The remainder of the division of 7 by 3 is 1. Hence the remainder of the division of a power of 7 by 3 is 1, and the remainder of the division of $7^m + 7^k$ by 3 is 2.

However, no square number divided by 3 leaves the remainder 2 (squares of the form $(3t)^2$ leave the remainder 0, and squares of the form $(3t \pm 1)^2$ leave the remainder 1). Hence there is no square among the numbers $7^m + 7^k$.

Problem 28

(a) $x = 1$, $y = 2$, $z = 3$ provide a solution of the equation $x^3 + y^3 + z^3 = x^2y^2z^2$.

(b) To tackle the general case, suppose that $x \le y \le z$, and write the equation

$$x^3 + y^3 + z^3 = nx^2y^2z^2$$

in the form

$$z = nx^2y^2 - \frac{x^3 + y^3}{z^2}. \tag{23}$$

Since z, n, x and y are integers, $(x^3 + y^3)/z^2$ must also be an integer; hence

$$x^3 + y^3 \ge z^2 \tag{24}$$

Moreover, $x/z < 1$ and $y/z < 1$ combined with (23) lead to the inequality $z \ge nx^2y^2 - (x + y)$, or

$$z^2 \ge [nx^2y^2 - (x + y)]^2. \tag{25}$$

(24) and (25) together imply that

$$n^2x^4y^4 < 2nx^2y^2(x + y) + x^3 + y^3,$$

or, after dividing by nx^3y^3,

$$nxy < 2 \left(\frac{1}{x} + \frac{1}{y} \right) + \frac{1}{nx^3} + \frac{1}{ny^3}. \tag{26}$$

If $x \ge 2$, then the right-hand side of (26) is smaller than 3, while the left-hand side is not less than 4. This contradiction shows that $x = 1$.

By substituting 1 for x into (26), the inequality reduces to

$$ny < 2 + \frac{2}{y} + \frac{1}{n} + \frac{1}{ny^3}. \tag{27}$$

For $y \geq 4$ the right-hand side of (27) is less than 4. Hence $y \leq 3$.

We saw previously that $(x^3 + y^3)/z^2$ must be an integer. Since $x = 1$, z^2 must divide $1 + y^3$. This condition, together with $z \geq y$ and $y \leq 3$, leads to the triples

1. $x = 1, y = 1, z = 1$
2. $x = 1, y = 2, z = 3$.

Both triples are solution sets for $x^3 + y^3 + z^3 = nx^2y^2z^2$; in case (1) $n = 3$ and in case (2) $n = 1$.

Thus, 1 and 3 are the only values of n for which the above equation has solutions in positive integers.

Problem 29

There are only finitely many ways to represent 50 as the sum of natural numbers; this implies that among the corresponding products there is at least one with maximum value. Denote the maximum value of the product by P, and the summands of 50 whose product is P by $s_1, s_2, \ldots, s_k$.

The following can be proved:

1. At least two of the summands are equal to 5. Namely, since 100 divides P, 5 divides P, so one of the summands, say $s_1 = 5m$. Suppose that $m > 1$. In that case $5m$ can be split into three positive summands: $5 + m + (4m - 5)$, whose product $5m(4m - 5)$ is greater than $5m$. But then the numbers 5, m, $4m - 5, s_2, s_3, \ldots, s_k$ would have a product greater than the largest possible product P. This contradiction rules out the case $m > 1$. Hence $s_1 = 5$.

Because 100 is divisible by 25, there is another summand equal to 5, say s_2.

2. There is either a summand s_3 equal to 4, or at least two summands s_3 and s_4 are equal to 2. To see this, note that P is an even number (100 divides P), so there is a summand, say s_3, equal to $2k$. If $k > 2$, then from $s_3 = 2 + k + (k - 2)$ it follows that $2k(k - 2)s_1 s_2 s_4 \cdots s_k > P$. Since this cannot be true, $k = 1$ or $k = 2$.

If $k = 2$ then $s_3 = 4$. If $k = 1$, then $s_3 = 2$, and there must exist a further summand $s_4 = 2$; this is because P is divisible by 100, which is a multiple of 4.

3. Having determined the summands with sum equal to $5 + 5 + 4 = 14$, it remains to find the summands of $50 - 14 = 36$ such that the product formed by these summands is a maximum.

Suppose that a summand $s_i > 4$. In that case, $s_i = 2 + (s_i - 2)$ and $2(s_i - 2) > s_i$. As before, this leads to contradiction.

No summand can be equal to 1. In fact, suppose that $s_i = 1$; then, for any s_j such that $j \neq i$ the sum $s_i + s_j$ is equal to $1 + s_j$ and $s_i s_j = 1 \cdot s_j < 1 + s_j$.

Thus by replacing the two summands s_i, s_j by their sum a greater product would be obtained.

This shows that the summands of 36 are all equal to 3, 2, or 4.

4. Suppose that there are more than 2 summands equal to 2. In that case $2 + 2 + 2$ can be replaced by $3 + 3$ and the product $3 \cdot 3$ is greater than $2 \cdot 2 \cdot 2$. It follows that there can be at most 2 summands equal to 2 in the set forming the maximum product.

Similarly, it can be shown that if one of the summands is 4, then there can be no more summands equal to 4 or 2.

5. We must check which of the following representations of 36 are possible:

$$36 = 2 + 2 + 3 + 3 + \cdots + 3 = 4 + 3 + 3 + 3 + \cdots + 3$$

or

$$36 = 2 + 3 + 3 + 3 + \cdots + 3$$

or

$$36 = 3 + 3 + 3 + \cdots + 3.$$

Obviously, only the last case can occur.

This proves that a representation of 50 as the sum of natural numbers with the greatest possible product divisible by 100 is of the form

$$50 = 5 + 5 + 2 + 2 + 3 + 3 + \cdots + 3$$

or

$$50 = 5 + 5 + 4 + 3 + 3 + \cdots + 3$$

Thus the greatest possibility for the product is $2^2 \times 5^2 \times 3^{12}$ which is 53 144 100.

Problem 30

There are polyhedra satisfying the assumptions of Problem 30 for which there is no sphere S' passing through all their vertices. Here is an example:

Take a cube C and over each face as a base construct a pyramid with apex outside the cube such that all triangular faces of the pyramid form angles of 45° with the base.

The resulting solid P has 12 faces and 14 vertices, 8 of which are the

vertices of the cube and 6 are the apexes of the pyramids. The edges of the
cube are not among the edges of P.

   All edges of P are of equal length, and since their distances from the centre
O of the cube are all the same, they all touch a sphere S with centre O.
However, the vertices of P do not lie on a common sphere, because the
sphere containing the eight vertices of C does not pass through the remaining
six vertices of P.

Section 4: Generalizing given problems

Problem 31 (E)

On *EH* construct the point *I* such that *AI* is parallel to *BH*, and on *GF*
construct *J* such that *CJ* is parallel to *BH*. Produce *HB* to meet *AC* in *M* and
LK in *N*.

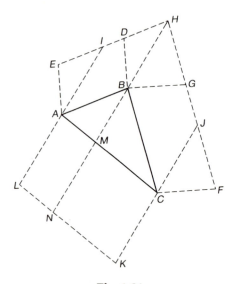

Fig. 1.61

   ABHI is a parallelogram having the same base *AB* and the same corresponding height as the parallelogram *ABDE*. On the other hand, *ABHI*
and *NMAL* are also parallelograms with equal bases (*BH* = *NM*) and equal
corresponding heights. Thus

$$\text{Area } AMNL = \text{Area } ABHI = \text{Area } ABDE.$$

Similarly,

$$\text{Area } MCKN = \text{Area } BCJH = \text{Area } BCFG.$$

This implies that

$$\text{Area } ACKL = \text{Area } ABDE + \text{Area } CBGF.$$

Problem 32 (E)

Denote the lengths of OA, OB and OC by a, b and c respectively. Since AOB, BOC and COA are right-angled triangles,

$$\left.\begin{array}{l} \text{Area } AOB = \tfrac{1}{2}ab \\[4pt] \text{Area } BOC = \tfrac{1}{2}bc \\[4pt] \text{Area } COA = \tfrac{1}{2}ac \end{array}\right\} \qquad (28)$$

The area of $\triangle ABC$ is $\tfrac{1}{2}AB \cdot CD$, where CD is the height of $\triangle ABC$ corresponding to AB.

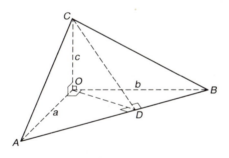

Fig. 1.62

It is well known that if O is the foot of the perpendicular from a point C outside a given plane AOB to that plane, and D is the foot of the perpendicular from O to a straight line AB in the plane AOB, then CD is perpendicular to AB (Prove it!). Thus $\triangle COD$ is a right-angled triangle. By Pythagoras' theorem, its hypotenuse can be expressed as $CD = \sqrt{(c^2 + OD^2)}$.

From $\tfrac{1}{2}OD \cdot AB = \text{Area } AOB = \tfrac{1}{2}ab$, it follows that

$$OD = \frac{ab}{AB} = \frac{ab}{\sqrt{(a^2 + b^2)}}$$

and

$$CD = \sqrt{\left(c^2 + \frac{a^2b^2}{a^2 + b^2}\right)}.$$

This implies that

$$\text{Area } ABC = \tfrac{1}{2}\sqrt{(a^2 + b^2)}\sqrt{\left(c^2 + \frac{a^2b^2}{a^2 + b^2}\right)}. \tag{29}$$

Combining (28) and (29) we find that

$$(\text{Area } AOB)^2 + (\text{Area } BOC)^2 + (\text{Area } COA)^2$$

$$= \tfrac{1}{4}(a^2b^2 + b^2c^2 + a^2c^2)$$

$$= \tfrac{1}{4}(a^2 + b^2)\left(c^2 + \frac{a^2b^2}{a^2 + b^2}\right)$$

$$= (\text{Area } ABC)^2.$$

Problem 33

We shall prove the statement in its general form, following the method described in [74], p. 38: First we prove an auxiliary statement:

(A) Of two pairs of numbers x, y and X, Y having the same sum, the pair with the greater product is the one whose numbers possess the smaller difference.

The proof of (A) is based on the following identities:

$$4xy = (x + y)^2 - (x - y)^2$$

and

$$4XY = (X + Y)^2 - (X - Y)^2.$$

Since $x + y = X + Y$, this implies that $xy > XY$ if and only if $|x - y| < |X - Y|$.

Let S be a sequence of positive numbers $a_1, a_2, \ldots, a_n$ with sum Σ. We distinguish two cases:

Case 1: $a_1 = a_2 = \cdots = a_n$. In that case $a_i = \Sigma/n$ for all $i = 1, 2, \ldots, n$, and the product $a_1 a_2 \ldots a_n$ is equal to $(\Sigma/n)^n$. Put $M = \Sigma/n$ and $\Pi = (\Sigma/n)^n$.

Case 2: $a_1, a_2, \ldots, a_n$ are not all equal. In that case at least one of the numbers, say a_1, must be greater than M, and at least one of the numbers, say

a_2, must be smaller than M. In the sequence S replace the terms a_1 and a_2 by $a_1' = M$ and $a_2' = (a_1 + a_2) - a_1'$; denote the new sequence $a_1', a_2', a_3, a_4, \ldots,$ a_n by S_1. In S_1 the sum of the terms is the same as in S, for $a_1' + a_2' = a_1 + a_2$. However, $|a_1' - a_2'| < |a_1 - a_2|$. Therefore, in view of (A), $a_1'a_2' > a_1 a_2$.

Thus the product P_1 of the terms in S_1 is greater than the product P of $a_1, a_2, \ldots, a_n$.

If the new numbers $a_1', a_2', a_3, \ldots, a_n$ are not all equal, the above process is iterated: a term $a_i > M$ is replaced by M, and a term $a_j < M$ by $(a_i + a_j) - M$. In the new sequence S_2 the sum of the terms is Σ and the product $P_2 > P_1$.

If we proceed in this fashion, then after a finite number k of steps, we obtain a sequence S_k with all terms equal to M whose product $P_n = M^n = \Pi$.

Since $P < P_1 < P_2 < \cdots < P_n = \Pi$, it follows that in Case 2 the product $a_1 a_2 \cdots a_n$ is less than Π.

This completes the proof of the general statement: 'The product of n positive numbers $a_1, a_2, \ldots, a_n$, whose sum is constant, attains its maximal value when $a_1 = a_2 = a_3 = \cdots = a_n$.

Problem 34

(a) Let O be the centre of the unit circle c and let v_i be the vector $\mathbf{OA}_i$ for $i = 1, \ldots, n$. Then the square of the distance $A_i A_j$ is the value of the scalar product

$$(\mathbf{v}_j - \mathbf{v}_i) \cdot (\mathbf{v}_j - \mathbf{v}_i) = (\mathbf{v}_j - \mathbf{v}_i)^2.$$

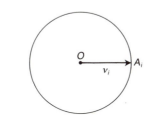

Fig. 1.63

Thus the sum S of the squares of all mutual distances can be expressed as

$$S = \tfrac{1}{2}[(\mathbf{v}_1 - \mathbf{v}_1)^2 + (\mathbf{v}_1 - \mathbf{v}_2)^2 + \cdots + (\mathbf{v}_1 - \mathbf{v}_n)^2$$
$$+ (\mathbf{v}_2 - \mathbf{v}_1)^2 + (\mathbf{v}_2 - \mathbf{v}_2)^2 + \cdots + (\mathbf{v}_2 - \mathbf{v}_n)^2$$
$$+ \cdots$$
$$+ (\mathbf{v}_n - \mathbf{v}_1)^2 + (\mathbf{v}_n - \mathbf{v}_2)^2 + \cdots + (\mathbf{v}_n - \mathbf{v}_n)^2]$$

$$= n(v_1^2 + v_2^2 + \cdots + v_n^2) - (v_1 + v_2 + \cdots + v_n)^2$$

$$= n^2 - (v_1 + v_2 + \cdots + v_n)^2.$$

It follows that S is always less than or equal to n^2. Equality is attained when the position vectors v_i of the points A_i all add up to $\mathbf{O}$.

(b) Clearly, the same method can be applied to points on the unit sphere and yields the same result.

Problem 35

(a) Introduce a Cartesian coordinate system in which the given points A_i have coordinates x_i, y_i for $i = 1, 2$ and the points P of the locus have coordinates x and y.

The condition $|PA_1|^2 + |PA_2|^2 = c$ is then transformed into the equation

$$(x - x_1)^2 + (y - y_1)^2 + (x - x_2)^2 + (y - y_2)^2 = c$$

which can be written in the form

$$\left[x - \frac{x_1 + x_2}{2} \right]^2 + \left[y - \frac{y_1 + y_2}{2} \right]^2 = C, \tag{30}$$

where C is a constant depending on c, x_1, y_1, x_2 and y_2.

Equation (30) represents a circle, a point, or the empty set, according as C is greater than, equal to or less than 0.

(b) Denote the coordinates of A_i by x_i, y_i and the coordinates of P by x, y. Then $|PA_i|^2 = (x - x_i)^2 + (y - y_i)^2$ and the equation

$$\alpha_1 |PA_1|^2 + \alpha_2 |PA_2|^2 + \cdots + \alpha_n |PA_n|^2 = \beta$$

can be rearranged in the form

$$dx^2 + dy^2 + ax + by + c = 0,$$

where a, b, c are determined by x_i, y_i, α_i and β, and $d = \alpha_1 + \alpha_2 + \cdots + \alpha_n$.

We distinguish two cases:

Case 1: $d \neq 0$. In this case the set of points P is a circle (if $(a^2 + b^2 - 4dc)/4d^2 > 0$), consists of a single element (if $(a^2 + b^2 - 4dc)/4d^2 = 0$), or is empty (if $(a^2 + b^2 - 4dc)/4d^2 < 0$).

Case 2: $d = 0$. Then the equation of the locus reduces to

$$ax + by + c = 0,$$

which represents a straight line (if at least one of a and b is not zero), the whole plane (if $a = b = c = 0$), or the empty set (if $a = b = 0$, $c \neq 0$).

Problem 36

The following solution is given by Sierpiński [92]. (For a different approach see Problem 53).

If $n = 2$, the equation $x_1^2 + x_2^2 = b_2^2$ has the integer solutions $x_1 = 3$, $x_2 = 4$ and $b_2 = 5$. b_2 is an odd number.

Suppose that for some n the equation

$$x_1^2 + x_2^2 + \cdots + x_n^2 = y_n^2 \tag{31}$$

has integer solutions $x_i = a_i$ and $y_n = b_n$ such that b_n is an odd number $2k + 1$.

Notice that $(2k + 1)^2 + (2k^2 + 2k)^2 = (2k^2 + 2k + 1)^2$. Consequently by adding $(2k^2 + 2k)^2$ to both sides of the identity

$$a_1^2 + a_2^2 + \cdots + a_n^2 = (2k + 1)^2$$

this is transformed into

$$a_1^2 + a_2^2 + \cdots + a_n^2 + (2k^2 + 2k)^2 = (2k^2 + 2k + 1)^2.$$

In other words, $x_1 = a_1$, $x_2 = a_2$, . . ., $x_n = a_n$, $x_{n+1} = 2k^2 + 2k$ and $y_{n+1} = 2k^2 + 2k + 1$ are solutions of the equation

$$x_1^2 + x_2^2 + \cdots + x_{n+1}^2 = y_{n+1}^2.$$

Thus equation (31) has integer solutions for all $n = 2, 3, \ldots$.

Problem 37

(a) Without loss of generality it can be assumed that none of the straight lines is 'horizontal'. (Otherwise turn the diagram.)

In order that the number of the regions is the greatest possible, any two lines should meet and no three of the n lines should be concurrent.

The regions into which such straight lines divide the plane are of two types:

(I) Regions which are bounded from 'below'; such regions have a 'lowest'

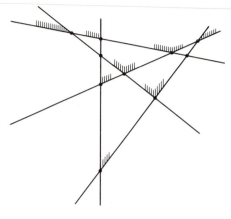

Fig. 1.64

point which is the intersection of two of the given lines. Conversely, every intersection of two lines from the given set is the 'lowest' point for exactly one region (Fig. 1.64).

Thus the number of regions of type (I) is the number of points determined by pairs of n lines no three of which are concurrent. This number is $C_2^n = n(n - 1)/2$.

(II) Regions which are not bounded from 'below'. To find their number draw a straight line ℓ across the plane such that all regions of type (I) are 'above' it. This is possible since there are finitely many regions of type I. The n lines meet ℓ in n points, dividing it into $n + 1$ parts. Each region of type (II) contains exactly one part of ℓ, and each part of ℓ is contained in a region of type (II). Thus there are $n + 1 = C_1^n + C_0^n$ regions of type (II).

It follows that the greatest number of regions into which n straight lines can divide the plane is $C_2^n + C_1^n + C_0^n$.

(b) The following is a generalization of the above problem in space:

n planes are given in arbitrary position. Find the maximum number of regions into which the planes can divide the space.

Solution: Without loss of generality it can be assumed that none of the plane is 'horizontal'.

The n planes will determine the maximum possible number of regions when any three planes meet in exactly one point and when no four planes have a point in common. In this way the n planes determine

$$C_3^n = \frac{n(n - 1)(n - 2)}{3 \cdot 2 \cdot 1}$$

points, each of which is the 'lowest' point of a region of type (I), that is, of a region bounded from 'below'.

Place a plane π 'below' all regions of type I. π cuts the n given planes in n straight lines which divide π into planar regions. According to (a), the number of these planar regions is $C_2^n + C_1^n + C_0^n$. It is easy to see that this must be the number of regions of type (II), that is, regions not bounded from 'below'.

It follows that the maximum number of regions in space is

$$C_3^n + C_2^n + C_1^n + C_0^n.$$

Problem 38

Denote the entry in the ith row and jth column of M_n^2 by m_{ij}.

(a) If $i = j$, then

$$m_{ii} = 1^2 + 2^2 + \cdots + (n-1)^2.,$$

Thus: $m_{11} \equiv m_{22} \equiv \cdots \equiv m_{nn}$ (modulo n).

(b) If $i \neq j$, put $j = i + k$, where $k \in \{\pm 1, \pm 2, \ldots, \pm(n-1)\}$. We have

$$m_{ij} \equiv (i-1)(i-1+k) + i(i+k) + \cdots$$

$$+ (n-1)(n-1+k) + 0 \cdot k + 1(1+k) + \cdots + (i-2)(i-2+k)$$

$$\equiv (i-1)^2 + i^2 + (i+1)^2 + \cdots + (n-1)^2 + 0^2 + 1^2 +$$

$$\cdots + (i-2)^2 + nik + \frac{kn(n-1)}{2} \text{(modulo } n)$$

Hence

$$m_{ij} \equiv m_{ii} + k\,\frac{n(n-1)}{2} \text{ (modulo } n).$$

We distinguish two cases:

Case 1: n odd. In this case $kn(n-1)/2$ is divisible by n. Therefore $m_{ij} \equiv m_{ii}$ (modulo n).

Then all entries of M_n^* are equal.

Case 2: n even

Here $m_{ij} \equiv m_{ii}$ (modulo n) exactly when $kn(n-1)/2 \equiv 0$ (modulo n. But

$$\frac{kn(n-1)}{2} \equiv 0 \text{ (modulo } n) \qquad \text{if } k = 2\ell \text{ for some } \ell,$$

and

$$\frac{kn(n-1)}{2} \equiv \frac{n(n-1)}{2} \text{ (modulo } n) \qquad \text{if } k = 2\ell + 1.$$

Hence M_n^* has two types of entries.

Problem 39

The $n \times n \times n$ chessboard $ABCDEFGH$ consists of n^3 cubic cells. Call a 'layer' any set of n^2 cells of the board forming an $n \times n \times 1$ cuboid. Let R be a set of r rooks governing all cells of the board.

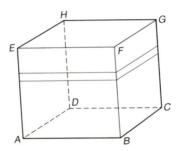

Fig. 1.65

Choose a layer L with the smallest number m of rooks. Suppose that L is 'horizontal', that is, parallel to $ABCD$. The rooks on L govern a certain number of 'rows' (parallel to AB), say m_1, and a certain number of 'columns' (parallel to AD), say m_2.

This leaves $(m - m_1)(m - m_2)$ cells on L not governed by rooks of L. Each of these cells must be governed by a rook in the 'vertical' direction (parallel to AE).

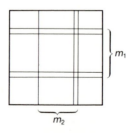

Fig. 1.66

Now consider the distribution of the rooks of R on all layers parallel to $ABFE$. These layers can be divided into two types:

Type I: Layers containing no rook standing on L. There are $n - m_1$ layers of type I; each of them contains at least $n - m_2$ rooks. Thus all layers of type I contain at least $(n - m_1)(n - m_2)$ rooks.

Type II: The remaining m_1 layers parallel to $ABFE$. Since m is the smallest number of rooks on all layers, the number of rooks on each such layer is at least m.

Hence all layers of type II contain at least $m_1 m$ rooks.

The above investigations imply that

$$r \geq (n - m_1)(n - m_2) + m_1 m.$$

Suppose that $m_1 \geq m_2$; then

$$r \geq (n - m_1)^2 + m_1^2 = \frac{n^2}{2} + \frac{(n - 2m_1)^2}{2}. \tag{32}$$

r must be an integer. Hence, in view of (32):

if n is even then the smallest possible value of r is $\dfrac{n^2}{2}$;

if n is odd then the smallest possible value of r is $\dfrac{n^2 + 1}{2}$.

Remark: In both cases the smallest possible value for r can be achieved. This is demonstrated by examples 1 and 2. The diagrams show the board from 'above'; the number k in the ith row and jth column indicates that the cell in the ith row and jth column of the kth horizontal layer is occupied by a rook (Fig. 1.67).

			7	4	5	6
			6	7	4	5
			5	6	7	4
			4	5	6	7
3	1	2				
2	3	1				
1	2	3				

Example 1

			8	5	6	7
			7	8	5	6
			6	7	8	5
			5	6	7	8
4	1	2	3			
3	4	1	2			
2	3	4	1			
1	2	3	4			

Example 2

Fig. 1.67

Problem 40

Introduce a coordinate system such that the centres of the cells of the board have coordinates (x, y, z) where x, y, z are non-negative integers. Place the rook at the origin O $(0, 0, 0)$.

Call a step on the board a path from the centre of one cell to the centre of the neighbouring cell. (Two cells are neighbours if they share a face.)

Any path from O to $P(i, j, k)$ consists of $i + j + k$ steps, of which i have to be parallel to the x-axis, j to the y-axis and k to the z-axis. Thus the number of different paths leading from O to P is the number of permutations of $i + j + k$ steps, of which i are of one kind, j of a second kind and k of a third kind. This number is well known to be equal to

$$\frac{(i + j + k)!}{i!\,j!\,k!}.$$

The formula

$$\frac{(i + j + k)!}{i!\,j!\,k!} = \frac{(i + j + k - 1)!}{(i - 1)!\,j!\,k!} + \frac{(i + j + k - 1)!}{i!\,(j - 1)!\,k!} + \frac{(i + j + k - 1)!}{i!\,j!\,(k - 1)!}$$

accords with the fact that the field with centre (i, j, k) can be approached from the three neighbouring fields with centres $(i - 1, j, k)$, $(i, j - 1, k)$ and $(i, j, k - 1)$.

The number pattern obtained on the board can be built up in the shape of a pyramid as follows.

Cut the board by planes π_n through the points $A_n(n, 0, 0)$, $B_n(0, n, 0)$ and $C_n(0, 0, n)$ for $n = 0, 1, 2, 3, \ldots$. Each plane contains those points (i, j, k) for which $i + j + k = n$. Label the point $P(i, j, k)$ with the number of paths the rook can take from O to P. The labels lying on π_n make a number triangle, forming the nth layer of the number pyramid P (Fig. 1.68).

The numbers $(i + j + k)!/(i!\,j!\,k!)$ are the coefficients in the expansion of the $(i + j + k)$th power of the trinomial $x + y + z$; therefore they are called trinomial coefficients.

Remark: In each triangle of the pyramid P add the numbers along the dotted lines. The resulting number sequences are

$$1;\ 1,1;\ 1,2,3,2,1;\ 1,3,6,7,6,3,1;$$
$$1,4,10,16,19,16,10,4,1;\ \ldots$$

They are the rows in the number triangle of Problem 11(c).

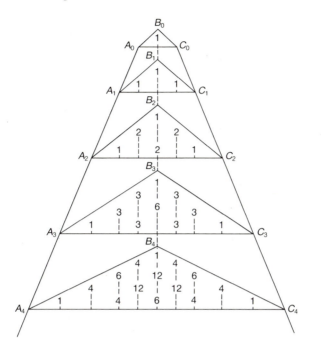

Fig. 1.68

Section 5: Converse problems

Problem 41 (E)

Let $\angle CAB = 2x$ and $\angle CBA = 2y$.

(a) If $AC = BC$, then $x = y$. Consequently, $\triangle ABO$ is isosceles and triangles ABA' and ABB' are congruent. Thus $AO = BO$ and $AA' = BB'$. Hence $OA' = OB'$.

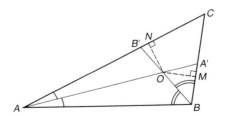

Fig. 1.69

But the converse statement does not hold:

(b) If $OA' = OB'$, then $\triangle ABC$ need *not* be isosceles, as shown below.

Drop the perpendiculars OM to BC and ON to AC. Since O is the common point of the bisectors of the angles in $\triangle ABC$, the distances OM and ON are equal. Therefore, the right-angled triangles OMA' and ONB' are congruent, and

$$\angle OA'M = \angle OB'N. \tag{33}$$

Four cases can be distinguished:

Case 1: Both M and N are outside the quadrilateral $CB'OA'$;
Case 2: Both M and N are inside the quadrilateral $CB'OA'$;
Case 3: Exactly one of M and N is inside the quadrilateral $CB'OA'$;
Case 4: $M = A'$ and $N = B'$.

In *Case 1* $\angle OA'M = 180° - (x+2y)$ and $\angle OB'N = 180° - (2x+y)$; thus, in view of (33), the angles x and y are equal and $\triangle ABC$ is isosceles.

In *Case 2* $\angle OA'M = x + 2y$ and $\angle OB'N = 2x + y$. This together with (33) implies that $x = y$; $\triangle ABC$ is isosceles.

In *Case 3* suppose that M is outside and N inside the quadrilateral $CB'OA'$. Then $\angle OA'M = 180° - (x+2y)$, while $\angle OB'N = 2x + y$. Hence $x + y = 60°$ and $\angle ACB = 180° - (2x+2y) = 60°$. The sides AC and BC are unequal.

If N is outside and M inside $CB'OA'$, then the same conclusion follows.

In *Case 4* the bisectors AA' and BB' are also altitudes in $\triangle ABC$. It follows that $\triangle ABC$ is equilateral.

We have proved that if $OA' = OB'$, then in $\triangle ABC$ the sides AC and BC are equal, or the angle at C is 60°.

Problem 42 (E)

(a) Place the square S in the coordinate system so that its vertices are the

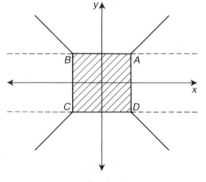

Fig. 1.70

points $A(a, a)$, $B(-a, a)$, $C(-a, -a)$ and $D(a, -a)$ (Fig. 1.70). Let L be the locus of points in the plane for which the sum s_x of the distances from BC and AD is equal to s_y, the sum of the distances from BA and CD.

Clearly, any point $P(x, y)$ between the straight lines DC and AB or between DA and CB is in L if and only if P belongs to S.

For a point P with $x \geqslant a$, $y \geqslant a$ the condition $s_x = s_y$ implies that

$$(x - a) + (x + a) = (y - a) + (y + a).$$

Thus $x = y$. Since the x- and y-axes are axes of symmetry for S, it follows that L consists of the points of S and of the points on the lines $y = x$ and $y = -x$.

(b) Let R be a rectangle with vertices $A(a, b)$, $B(-a, b)$, $C(-a, -b)$ and $D(a, -b)$ such that $a > b$ (Fig. 1.71).

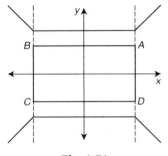

Fig. 1.71

In this case no point between the parallel lines AB and CD belongs to L.

If $P(x, y)$ is in L such that $-a \leq x \leq a$, $y > 0$, then $s_x = 2a$ and $s_y = (y - b) + (y + b)$. Thus $s_x = s_y$ implies that $y = a$.

If $P(x, y)$ is in L such that $x > a$, $y > a$, then $s_x = (x - a) + (x + a)$ and $s_y = (y - b) + (y + b)$. Hence $s_x = s_y$ leads to $y = x$.

Considerations of symmetry imply that the set L consists of the points $P(x, y)$ on the lines $y = \pm a$ such that $|x| \leq a$, and of the points on the lines $y = \pm x$ such that $|x| \geq a$.

Problem 43

(a) Suppose that the pentagon $P = A_1 A_2 A_3 A_4 A_5$ has been constructed. Let M_1, M_2, M_3, M_4, M_5 be the midpoints of the consecutive sides $A_1 A_2$, $A_2 A_3$, ..., $A_5 A_1$.

Choose any point B_1 in the plane not on the perimeter of P. Rotate B_1 through $180°$ about M_1 to obtain B_2; rotate B_2 about M_2 to obtain B_3, B_3

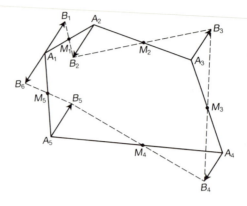

Fig. 1.72

about M_3 to obtain B_4, and so on. The vectors $\mathbf{A}_i\mathbf{B}_i$ satisfy the following relations:

$$\mathbf{A}_1\mathbf{B}_1 = -\mathbf{A}_2\mathbf{B}_2 = \mathbf{A}_3\mathbf{B}_3 = -\mathbf{A}_4\mathbf{B}_4 = \mathbf{A}_5\mathbf{B}_5 = -\mathbf{A}_1\mathbf{B}_6.$$

Thus B_6, A_1 and B_1 are collinear, and A_1 is the midpoint of the straight line segment B_6B_1.

Therefore, to construct P when P' is given, choose any point B_1 not on the perimeter of P, construct successively the points B_2, B_3, B_4, B_5 and B_6, and then the midpoint A_1 of B_1B_6. Then A_2, A_3, A_4 and A_5 are constructed by rotating the previously obtained vertex A_i about M_i through $180°$ for $i = 1$, 2, 3, 4.

(b) The same construction method applies to any n-gon where n is an *odd* number; in this case from P' a unique polygon P is obtained.

If n is *even*, then $\mathbf{A}_1\mathbf{B}_1 = \mathbf{A}_1\mathbf{B}_{n+1}$. Hence the problem has no solution if $B_{n+1} \neq B_1$. If $B_{n+1} = B_1$, there are infinitely many solutions since A_1 can be taken in an arbitrary position.

Problem 44 (E)

(a) If $t = n(n+1)/2$, then

$$8t + 1 = 8\frac{n(n + 1)}{2} + 1 = 4n^2 + 4n + 1 = (2n + 1)^2.$$

(b) If $8t + 1 = m^2$, then m must be an odd number of the form $m = 2k + 1$. Thus

$$8t + 1 = (2k + 1)^2$$

and

$$t = \frac{(2k + 1)^2 - 1}{8} = \frac{4k^2 + 4k}{8} = \frac{k(k + 1)}{2}.$$

Problem 45

(a) Fermat's little theorem can be proved by induction on a. For $a = 1$, $1^p - 1 = 0$, which is divisible by p. Suppose that $a^p - a$ is divisible by p for any natural number a. For $a + 1$ it follows that

$$(a+1)^p - (a+1) = \binom{p}{0}a^0 + \binom{p}{1}a^1 + \cdots + \binom{p}{p}a^p - a - 1$$

$$= \underbrace{(a^p - a)}_{} + \underbrace{\left[\binom{p}{1}a + \binom{p}{2}a^2 + \cdots + \binom{p}{p-1}a^{p-1}\right]}_{}$$

$$= \quad A \quad + \qquad\qquad\qquad\qquad B$$

Every summand in B is divisible by p, so B is divisible by p. On the other hand, A is divisible by p by the induction hypothesis.

Hence $(a + 1)^p - (a + 1)$ is divisible by p. This proves (a).

(b) $341 = 31 \times 11$, hence 341 is a composite number. Moreover,

$$2^{341} - 2 = (2^{31})^{11} - 2 = \underbrace{(2^{31})^{11} - 2^{11}}_{A} + \underbrace{2^{11} - 2}_{B}.$$

$$A = 2^{11}(2^{30 \cdot 11} - 1) = 2^{11}[(2^{10})^{33} - 1];$$

hence A is divisible by $2^{10} - 1$.

$$B = 2(2^{10} - 1).$$

Hence B is also divisible by $2^{10} - 1$.

Since $2^{10} - 1 = 3 \cdot 341$, both A and B are divisible by 341. Thus

$$341 \text{ divides } 2^{31 \times 11} - 2.$$

(c) Suppose that

$$n \text{ is an odd composite number dividing } 2^n - 2. \qquad (34)$$

Our aim is to show that $m = 2^n - 1$ is a composite number dividing $2^m - 2$.

Since n is composite, it can be written in the form $n = ab$, where a, b are integers and $1 < a < n$. Thus $m = 2^{ab} - 1$, which is divisible by $2^a - 1$.

Hence m is an odd composite number.

According to (34) the odd number n divides $2^n - 2$; hence n divides $2^{n-1} - 1$. Put $2^{n-1} - 1 = kn$. From

$$2^{m-1} = 2^{2^n - 2} = 2^{2(2^{n-1} - 1)} = 2^{2kn} = (2^n)^{2k}$$

it follows that

$$2^{m-1} - 1 = (2^n)^{2k} - 1$$

which is divisible by $2^n - 1$.

Thus $2(2^{m-1} - 1) = 2^m - 2$ is also divisible by $2^n - 1$, that is by m.

Problem 46

(a) If x and y are two natural numbers, and a and b two integers such that

$$ax + by = 1, \tag{35}$$

then any common divisor d of x and y must divide the right-hand side of (35), that is 1. Hence $d = 1$; x and y are relatively prime.

(b) Suppose that a and b are relatively prime. Consider the set $S = \{0, 1, 2, \ldots, b-1\}$ of the remainders of natural numbers after division by b, and form the products ai, $i \in S$. Modulo b these products are different for different values of i; otherwise, for $i > j$, $ai - aj = a(i - j)$ would be divisible by b. Since $i - j < b$, and b and a are relatively prime, this cannot happen. Thus for a certain value of i, say x, the remainder of ax after division by b is 1. That is, there exists an integer y such that

$$ax + by = 1.$$

Problem 47

(a) The first triple $x = 1$, $y = 1$, $z = 1$ can be spotted as a solution of

$$x^2 + y^2 + z^2 = 3xyz \tag{36}$$

by trial and error.

Solutions of (36), different from (1,1,1), are shown in the tree diagram (Fig. 1.73).

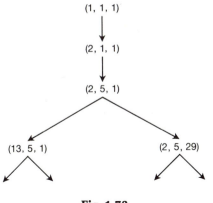

Fig. 1.73

To find the successor of $(1,1,1)$ in the diagram substitute $y = 1$ and $z = 1$ into (36). The quadratic equation

$$x^2 + 1 + 1 = 3x$$

has a second solution, $x = 2$, different from $x = 1$. Thus the triple $(2,1,1)$ is a set of solutions for (36).

To form the next successor in the diagram, put $x = 2$, $z = 1$ into (36). The equation

$$4 + y^2 + 1 = 6y$$

has the solution $y = 5$, different from the solution $y = 1$ in the previous triple. Hence $(2,5,1)$ is a new triple.

From $(2,5,1)$ two new triples can be deduced, corresponding to the equations

$$x^2 + 5^2 + 1^2 = 3 \cdot 5 \cdot 1 \cdot x \qquad \text{and} \qquad 2^2 + 5^2 + z^2 = 3 \cdot 2 \cdot 5 \cdot z.$$

Proceeding in the same way, solutions of (36) can be constructed step by step. (The equation (36) has infinitely many solutions in integers.)

(b) The investigation will be carried out in a number of steps.

Step 1:   If two of the numbers in the solution (x, y, z) are equal, say $x = y$, then $x = y = 1$, and $z = 1$ or $z = 2$.

For, let $x = y$, say. Then (36) reduces to

$$2x^2 + z^2 = 3x^2 z,$$

implying that z^2 is divisible by x^2. Putting $z^2 = a^2 x^2$ leads to $2 + a^2 = 3ax$.
   Thus $2 = a(3x - a)$. This can happen only if $a = 1$, when $x = y = z = 1$, or if $a = 2$ when $z = 2x = 2y = 2$.

Step 2:   Let $t = (a, b, c)$ be a triple of natural numbers satisfying (36) such that $a > b > c$. From t three 'adjacent triples' can be obtained satisfying (36):

$$t_1 = (a', b, c), \qquad t' = (a, b', c) \qquad \text{and} \qquad t'' = (a, b, c').$$

We shall show that:

1.   in t' and t'' the maximal elements are greater than a, and
2.   in t_1 the maximal element is less than a.

   To prove (1) recall that b and b' are the solutions of the quadratic equation $y^2 - 3acy + a^2 + c^2 = 0$, so that $b + b' = 3ac$. This relation leads to $b' = 3ac - b > a$. Similarly, one shows that $c' > a$.
   (2) a and a', being the solutions of $x^2 - 3bcx + b^2 + c^2 = 0$, are of the form

$$x_{1,2} = \frac{3bc}{2} \pm \sqrt{\left[\left(\frac{3bc}{2}\right)^2 - b^2 - c^2\right]}.$$

It is easy to verify that

$$x_2 = \frac{3bc}{2} - \sqrt{\left[\left(\frac{3bc}{2}\right)^2 - b^2 - c^2\right]}$$

is smaller than b. Thus $x_2 = a'$. The maximal element in t_1 is b, which is smaller than a.

Step 3:   According to Step 2, from any triple t of distinct natural numbers a, b, c satisfying (36) a triple t_1 is obtained in which the maximal element is smaller than the maximal element in t. Unless some elements of t_1 are equal, from t_1 a new triple t_2 with maximal element less than that of t_1 is constructed. This process of constructing triples with decreasing maximal elements cannot be continued indefinitely, since the maximal element are all non-negative integers. Thus, after a certain number of steps, a triple t_k is obtained in which at least two elements must be equal.
   According to Step 1 it is either $t_k = (1,1,1)$ or t_k is one of the triples $(2,1,1)$,

(1,2,1) or (1,1,2). If $t_k = (1,2,1)$ or $(1,1,2)$, permute its elements to form the triple $(2,1,1)$.

The triples $(1,1,1)$ and $(2,1,1)$ are in the tree diagram. By reversing each stage in the construction of t_k from t, one verifies that t, or one of its permutations, belongs to the tree diagram.

Remark: The numbers in the tree diagram are known in the literature as the 'Markoff numbers' (see e.g. [60]).

Problem 48

(a) Let π_1 be the plane containing $\Delta A_1 B_1 C_1$ and π_2 the plane containing $\Delta A_2 B_2 C_2$. The points P, Q, R belong to both π_1 and π_2. Since all points common to two non-parallel planes are on a straight line, P, Q and R are collinear.

(b) The converse of Desargues' theorem in space is the following statement:

Let $A_1 B_1 C_1$ and $A_2 B_2 C_2$ be two triangles in two non-parallel planes such that the three pairs of straight lines: $A_1 B_1$ and $A_2 B_2$, $B_1 C_1$ and $B_2 C_2$ and $C_1 A_1$ and $C_2 A_2$ meet in three (necessarily collinear) points P, Q and R respectively. Then the straight lines $A_1 A_2$, $B_1 B_2$ and $C_1 C_2$ are either all parallel to one another, or all meet in a point S.

Proof. $A_1 B_1$ and $A_2 B_2$ belong to some plane γ. $B_1 C_1$ and $B_2 C_2$ belong to a plane α, and $C_1 A_1$ and $C_2 A_2$ belong to a plane β. These three planes intersect pairwise in three lines: α and β in $C_1 C_2$, β and γ in $A_1 A_2$, and γ and α in $B_1 B_2$.

Three distinct lines along which three planes intersect pairwise are either all parallel or else they meet in some point S.

This proves our statement.

Remark. Problem 59 deals with Desargues' theorem in the plane.

Problem 49

(a) Let r be the radius of the circle σ inscribed in a triangle ABC with sides $a = 3, b = 4, c = 5$. The area of ΔABC is equal to

$$\mathbb{A} = \tfrac{1}{2}r(a + b + c) = \tfrac{1}{2}r \cdot 12 = 6r.$$

On the other hand, the triangle with sides 3, 4, 5 is right-angled, hence $\mathbb{A} = \tfrac{1}{2}ab = 6$.

It follows that $r = 1$.

(b) Suppose that $r = 1$.

Denote the points where σ touches the sides a, b, c by A_1, B_1, and C_1 respectively. Put $x = AB_1 = AC_1, y = BC_1 = BA_1$ and $z = CA_1 = CB_1$. Then the area $\mathbb{A}$ of ΔABC is

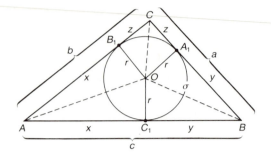

Fig. 1.74

$$\mathbb{A} = \tfrac{1}{2}r(a + b + c) = x + y + z. \tag{37}$$

On the other hand, applying Heron's formula:

$$\mathbb{A} = \sqrt{[s(s - a)(s - b)(s - c)]}$$

where $2s = a + b + c = 2x + 2y + 2c$, leads to

$$\mathbb{A} = \sqrt{[(x + y + z)xyz]}. \tag{38}$$

Combining (37) and (38) one obtains

$$x + y + z = xyz. \tag{39}$$

There are two possibilities:

(a) $2s$ is odd.   Since $2x = 2s - 2a$, $2y = 2s - 2b$, $2z = 2s - 2c$, it follows that $2x$, $2y$ and $2z$ are all odd numbers, and so is the product $8xyz$. However, in view of (39),

$$8xyz = 4(2x + 2y + 2z)$$

and the right-hand side of the above equality represents an even number.

This contradiction rules out the possibility that $2s$ is odd.

(b) $2s$ is even.   In this case x, y and z are all integers. Let $x \geq y \geq z$. Dividing (39) by x gives

$$1 + \frac{y}{x} + \frac{z}{x} = yz.$$

From

$$1 < 1 + \frac{y}{x} + \frac{z}{x} < 4$$

it follows that $yz = 3$ or $yz = 2$.

If $yz = 3$, then $y = x = z$ and $x^2 = 3$. This cannot happen, since x is an integer. If $yz = 2$, then $z = 1$, $y = 2$ and $x = 3$.

This proves that $a = 3$, $b = 4$ and $c = 5$.

Problem 50

(a) Let π be a plane meeting a sphere S in more than one point. Denote the centre of S by O and the foot of the perpendicular from O to π by O' (Fig. 1.75).

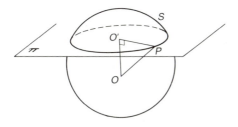

Fig. 1.75

If P is any point common to S and π, then from the right-angled triangle $OO'P$ it follows that

$$(O'P)^2 = (OP)^2 - (OO')^2 = R^2 - d^2,$$

where R is the radius of S and d the distance of O from π. Thus P is on the circle c with centre O' and radius $r = \sqrt{(R^2 - d^2)}$.

Conversely, let Q be any point on c; then its distance from O is given by $OQ = \sqrt{[(O'Q)^2 + (OO')^2]} = \sqrt{(r^2 + d^2)} = R$. Hence Q belongs to the intersection of π and S.

This shows that a plane meeting S in more than one point intersects it along a circle.

Let S be a surface such that any plane meeting S in more than one point cuts it along a circle.

Suppose that a plane π intersects S in a circle c. Denote by C the centre of c, and by p the straight line, perpendicular to π through C (Fig. 1.76).

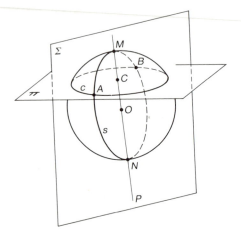

Fig. 1.76

Consider any plane Σ through p. Σ meets c in two distinct points A, B. These points are on S. Therefore, according to our assumptions, Σ cuts S in a circle s. The centre O of s is on the perpendicular bisector of AB, that is on p. The line p contains two points of s; these are the only points M and N of p on S.

By rotating Σ about p the circle s describes *a* sphere S^*. Clearly, all points of S^* belong to S.

Suppose that S contains a point K, not on S^*. Since K cannot belong to p, it determines together with p a plane Δ. The plane Δ passes through O. Hence it cuts S^* in a circle k. The circle k is in S; therefore the intersection of Δ with S contains k as well as K outside k. This contradicts the assumption of the problem.

Thus no point outside S^* belongs to S. The surface S is a sphere.

II Approaches to problem solving

Introduction

'Devising a plan, conceiving the idea of an appropriate action, is the main achievement in the solution of a problem', claims Polya in his famous book on problem solving, *How to Solve It*.

According to Polya, a good idea is a piece of good fortune and we have to deserve it by perseverance:

> An oak is not felled at one stroke. If at first you don't succeed, try, try again. It is not enough to try repeatedly. We must try different means, vary our trials.

Problem solving requires a versatile mind, but one cannot be versatile without a fair knowledge of techniques and methods of discovery. The aim of this chapter is to present a selection of approaches to problem solving, applied to problems in the corresponding sections.

There are eight sections in this chapter, concentrating on the following hints for problem solvers:

1. Express the problem in a 'different language'.
2. Extend the field of investigation.
3. Find out: Is some mathematical transformation involved in a given problem? Do any properties of the objects considered remain invariant under this transformation? If so, make use of the invariants.
4. Make use of extremal (minimal or maximal) elements.
5. Try the method of infinite descent.
6. Try mathematical induction.
7. Attempt proof by contradiction.
8. Employ physics.

Brief descriptions of the above approaches are given below.

1. *To express a problem in a 'different language'* means to rephrase the problem and to consider its equivalent in an appropriate branch of mathematics. This can greatly facilitate the solution. Here is an example:

Problem A ([20])

$[x]$ denotes the largest integer not greater than x. If p and q are relatively prime natural numbers, prove that

$$\left[\frac{p}{q}\right] + \left[\frac{2p}{q}\right] + \left[\frac{3p}{q}\right] + \cdots + \left[\frac{(q-1)p}{q}\right] = \frac{(p-1)(q-1)}{2}.$$

This number-theoretical problem appears to be difficult, until one realizes that the terms $[ip/q]$ have a neat interpretation in a two-dimensional coordinate system:

Since p and q are relatively prime, $[ip/q]$ is the number of points with integer coordinates (i, y), where $0 < y < ip/q$ for all $i = 1, 2, \ldots, q-1$. Hence the sum

$$s = \left[\frac{p}{q}\right] + \left[\frac{2p}{q}\right] + \cdots + \left[\frac{(q-1)p}{q}\right]$$

is the number P of points with integer coordinates inside the right-angled triangle OAB with vertices O $(0, 0)$, A $(q, 0)$ and B (q, p) (Fig. 1). In other words, Problem A is replaced by

Problem A'

Find P.

The solution to Problem A' is straightforward. The hypotenuse OB of $\triangle OAB$ contains no point with integer coordinates, since p and q are relatively prime. Thus P is half the number of points with integer coordinates inside the rectangle $OABC$, that is $P = \frac{1}{2}(p-1)(q-1)$.

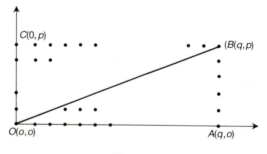

Fig. 2.1

Beginners are advised to work through Problems 51(E) and 52(E). Problem 53 offers a different approach to Problem 36.

2. *Extending the field of investigation* means studying a particular question in a wider context. For example, the proof of Desargues' theorem about triangles in a plane (Problem 59) can be substantially simplified by

introducing pyramids with vertices outside the given plane. (See solution to Problem 59.)

3. The following example explains what is meant by *using invariants of transformations*:

Problem B

A ladder of length 2 stands on a horizontal floor, leaning against a vertical wall. The ladder slides down. Describe the path of the ladder's midpoint.

During the ladder's motion the position of its midpoint is transformed into points M_t, depending on the time t after the start of the motion. The distance of M_t from O (Fig. 2.2) is equal to half the ladder's length. Since this length does not change, the length of OM_t is also an invariant of the transformation: $OM_t = \frac{1}{2}\ell$. Thus M_t describes an arc of the circle with centre O and radius $\ell/2$.

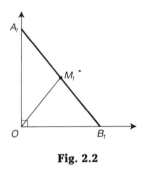

Fig. 2.2

A tip for beginners: Investigate Problems 60(E) and 61(E) by looking systematically at various situations which can arise. Careful analysis of the pattern may provide the correct answer. This work should be followed by studying the proofs in Part II.

4. *Making use of extremal elements* is illustrated by the following example:

Problem C

Each field of an infinite chessboard is occupied by a natural number. Each of these numbers is the average of the numbers in the four neighbouring fields (Fig. 2.3). Prove that all numbers on the board are equal.

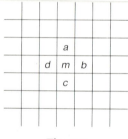

Fig. 2.3

Any set of natural numbers contains a smallest element. Thus, among the numbers on the board there is a minimal number m. If a, b, c, d are the numbers in the neighbouring fields, then

$$m = \frac{a + b + c + d}{4},$$

or

$$4m = a + b + c + d. \tag{1}$$

Since m is minimal, $a \geq m$, $b \geq m$, $c \geq m$ and $d \geq m$.

Suppose that at least one of a, b, c, d is greater than m. In that case

$$a + b + c + d > 4m,$$

contradicting (1). Thus $a = b = c = d = m$.

It is left to the reader to deduce that all numbers on the board must be equal to m.

5. *The method of infinite descent*, known already in ancient Greece, is a powerful tool for solving a wide range of problems. The method is especially suitable for proving negative statements, like the insolubility of an equation, or the impossibility of a construction. A typical way of applying the method is the following:

It is assumed that a given problem has a solution S. Starting with S, a never-ending sequence of solutions is constructed, although the nature of the problem indicates that any sequence of solutions must have a final term. This contradiction proves that the problem has no solution. (For proof by contradiction see also Section 7.)

An example is as follows:

Problem D

Find all pairs of positive integers x, y satisfying the equation

$$x^2 - 2y^2 = 0. \tag{2}$$

Solution

Suppose that there exist two positive integers a_1, b_1 such that

$$a_1^2 - 2b_1^2 = 0.$$

This implies that a_1 is an even number, that is, $a_1 = 2a_2$ for some positive integer a_2. From

$$(2a_2)^2 - 2b_1^2 = 0$$

it follows that

$$2a_2^2 - b_1^2 = 0.$$

Thus b_1 is even; it can be written as $b_1 = 2b_2$, with b_2 a positive integer. Substituting this into the last equation, we get that

$$2a_2^2 - (2b_2)^2 = 0,$$

or

$$a_2^2 - 2b_2^2 = 0. \tag{3}$$

This means that a_2, b_2 is another pair of solutions of equation (2). Since $a_1 = 2a_2$,

$$a_1 > a_2.$$

Moreover, the above equations imply that $a_1 > b_1 > a_2 > b_2$. (3) shows that a_2 is even, that is $a_2 = 2a_3$ for some positive integer a_3.

Repeating the above arguments, an infinite sequence of natural numbers is obtained, in which each term is smaller than the previous one:

$$a_1 > b_1 > a_2 > b_2 > a_3 > b_3 > \cdots.$$

However, every set of natural numbers has a smallest element, therefore the above infinite sequence cannot exist.

This contradiction shows that there are no positive integers satisfying (2).

Remark: The above arguments yield the proof of the well-known statement:

$$\sqrt{2} \text{ is an irrational number.}$$

Otherwise $\sqrt{2}$ would be a rational number p/q; this would imply that p and q are positive integers satisfying (2), which is impossible.

Beginners are advised to attempt Problems 64(E) and 65(E).

6. *The use of mathematical induction* is based on the following logical principle:

Let $\{P_n\}$ be a sequence of propositions, depending on the natural number n.
   If the following two conditions are satisfied:
1.   The first proposition is true, and
2.   there is a method for showing that *if* any proposition P_k is true, then P_{k+1} is *also* true,
then the proposition P_n is true for *all* natural numbers n.[1]

Thus to prove a statement P_n by mathematical induction means to verify conditions (1) and (2).

Apart from the two problems in this chapter we used induction in solutions of several problems in Chapter I. Notice that induction is involved in the method of infinite descent, described in item (6) above.

7. Instead of proving that a statement S is correct, it is often easier to show that the opposite of S is false. This implies that S is true. This method of proof is known as *proof by contradiction*. Problem D provides an example; another example is the following statement, well known from school geometry:

S: If a, b, c are three straigth lines in a plane such that a and b are parallel to c, then a is parallel to b.

To prove S, assume that the opposite is true:

S': a is not parallel to b.

Since a and b are in a plane, S' implies that a and b meet in a point P. In that case through P there are two lines parallel to c. From school geometry it is known that this cannot be the case, because

(E)   Through a given point P there is exactly one straight line parallel to a given straight line c.

Thus S' is false. Therefore S is true.

[1] Our 'logical principle' follows directly from one of Peano's axioms for the natural numbers (see e.g. [80]).

Remark: Statement (E) is accepted in the so-called Euclidean geometry as an unquestionable truth. It does not hold in non-Euclidean geometries. (See Chapter III, Section 4.)

8. Mathematical methods play an important role in the natural sciences (see recommended reading). At the same time science inspires mathematicians, often initiating remarkable discoveries (e.g. the discovery of quaternions; Chapter III, Section 3). The *interplay between mathematics and the sciences* — especially physics — is noticeable already in school curricula. For example:

The statement that 'The medians in an arbitrary triangle meet in a common point' can be interpreted as the consequence of the fact that each median carries the centre of gravity of the triangular plate *ABC*.

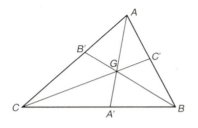

Fig. 2.4

Advanced problem solvers are advised to study the recommended reading (e.g. [40], [44], [45]). Beginners should investigate Problem 69(E).

Part I: Problems

Section 1: Expressing the problem in a different language

Problem 51 (E)

(a) A set of dominoes consists of all pieces carrying 0 to 6 dots on their halves. Is it possible to construct a chain out of all domino pieces of the set such that touching halves carry the same number of dots?

(b) All dominoes with 6 dots on their halves are removed. Is it possible to make a chain out of the remaining dominoes of the set such that touching halves carry the same number of dots?

Problem 52 (E)

(a) Prove that for any positive numbers $a_1, a_2, \ldots, a_n, b_1, b_2, \ldots, b_n$ the following relation holds:

$$\sqrt{(a_1^2 + b_1^2)} + \sqrt{(a_2^2 + b_2^2)} + \cdots + \sqrt{(a_n^2 + b_n^2)}$$
$$\geq \sqrt{[(a_1 + a_2 + \cdots + a_n)^2 + (b_1 + b_2 + \cdots + b_n)^2]}.$$

(b) When does equality hold?

Problem 53

Prove that for any natural number n the equation

$$a_1^2 + a_2^2 + \cdots + a_n^2 = b^2$$

has infinitely many solutions in positive integers.

Problem 54 (Kvant, M693, 1982, No. 3)

In an estate there are 1000 inhabitants. Every day each inhabitant tells the news he/she heard on the previous day to all his/her acquaintances. It is known that in this way eventually all news is transmitted to all inhabitants of the estate.

  Prove that it is possible to select 90 inhabitants such that if an item of news is given to them at the same time, then in 10 days this will be received by all inhabitants of the estate.

Problem 55

(a) Two points are chosen at random on a straight line segment AB, dividing it into three segments. What is the probability that a triangle can be constructed out of these three segments?
(b) Generalize this problem.

Section 2: Extending the field of investigation

Problem 56 (Kvant, M702, 1982, No. 5)

Denote by S_n the sum of the first n prime numbers (that is, $S_1 = 2, S_2 = 2 + 3 = 5, S_3 = 2 + 3 + 5 = 10$, etc.). Is it true that for any natural number n there is a square number between S_n and S_{n+1}?

Problem 57

Construct a set S of circles in the plane such that any circle of S touches exactly five circles of the set, all in different points.

Problem 58

Is the number $[(2 + \sqrt{3})^n]$ odd for all values of the natural number n? ($[x]$ denotes the greatest integer not exceeding x.)

Problem 59

Prove the theorem of Desargues in the plane:

Let ABC and $A'B'C'$ be two triangles in the same plane such that the straight lines AA', BB' and CC' meet in a common point S_1. If AB meets $A'B'$ in a point P, BC meets $B'C'$ in Q, and CA meets $C'A'$ in R, then P, Q and R lie on a common straight line.

Section 3: The use of invariants of transformations

Problem 60 (E) (Kvant, M913, 1985, No. 7)

There are 45 chameleons on an island: 17 of them are yellow, 15 are grey and 13 are blue. The chameleons wander around, meeting occasionally. At each meeting only two chameleons are present.

If two chameleons of the same colour meet, their colours remain unchanged. If two chameleons of different colours meet, both change their colour into the third colour (e.g. if a yellow chameleon meets a grey chameleon, they both change their colour into blue).

Could it happen that at a certain instant all chameleons on the island have the same colour?

Problem 61 (E) (Solved by Copernicus in the sixteenth century)

A circle c' rolls without slipping along the inside of a stationary circle c. The diameter of c' is half of the diameter of c. A point M is marked on the circumference of c'. Describe the path of M.

Section 4: The use of extremal elements

Problem 62

Does the equation

$$x^2 + y^2 = 3(z^2 + u^2)$$

have solutions in positive integers?

Problem 63 (E)

P is a set of n points in the plane such that any three points of P form a triangle of area at most 1.

Prove that all points of P are contained in a triangle of area not greater than 4.

Section 5: The method of infinite descent

Problem 64 (E)
Is it possible to cut a cube into a finite number of smaller cubes, all of different sizes?

Problem 65 (E)
A lattice point in a cartesian coordinate system is a point whose coordinates are integers.

(a) Is it possible to construct a regular pentagon in a two-dimensional coordinate system so that its vertices are lattice points?

(b) Generalize the above problem to other regular polygons in a two-dimensional coordinate system.

Section 6: Mathematical induction

Problem 66
Denote by $S_{k,n}$ the sum of the kth powers of the first n natural numbers for $k = 1, 2, 3, \ldots$.

$$S_{k,n} = 1^k + 2^k + \cdots + n^k.$$

Prove that $S_{k,n}$ is a polynomial in n of degree $k + 1$ with leading coefficient $1/(k + 1)$. (Thus
$$S_{k,n} = a_{k+1}n^{k+1} + a_k n^k + \cdots + a_2 n^2 + a_1 n + a_0,$$

where $a_{k+1} = 1/(k + 1)$.)

Problem 67
The vertices of a tetrahedron T are numbered 1, 2, 3 and 4. T is divided into a finite number of smaller tetrahedra T_i such that any two smaller tetrahedra have either just one face in common, or just an edge in common, or just a vertex in common, or no point in common. The vertices of all T_i are numbered by using some of the numbers 1, 2, 3, 4 in an arbitrary fashion. The only restrictions are that: Vertices of T_i which belong to some face f of T can be labelled only by numbers at the vertices of f, and vertices of T_i on some edge e of T can be labelled only by numbers at the endpoints of e.

Prove that there is at least one tetrahedron T_i whose vertices carry four different numbers 1, 2, 3 and 4.

Section 7: Proof by contradiction

Problem 68 (Bundeswettbewerb Mathematik 1973/74, 1st round [90])

Seven polygons of area 1 lie inside a square of side length 2. Prove that at least two of the polygons intersect in a region of area not less than $\frac{1}{7}$.

Section 8: Employing physics

Problem 69 (E)

(a) Construct a triangle of given base and area, having the smallest possible perimeter.
(b) In a given acute-angled triangle ABC inscribe a triangle of the smallest perimeter.

Problem 70 (Elemente der Mathematik, Vol. 13–15)

Each side of an arbitrary triangle ABC is divided into n equal parts. Denote the division points on BC by $A_1, A_2, \ldots, A_{n-1}$, on CA by $B_1, B_2, \ldots, B_{n-1}$ and on AB by $C_1, C_2, \ldots, C_{n-1}$.

(a) Find the smallest *odd* number n such that for some $i, j, k \in \{1, 2, \ldots, n-1\}$ the straight lines AA_i, BB_j and CC_k meet in a common point.
(b) Extend the above problem to three-dimensional space.

Part II: Solutions

Section 1: Expressing the problem in a different language

Problem 51 (E)

(a) A domino chain corresponds to an oriented graph as follows:

The vertices of the graph labelled 0, 1, 2, 3, 4, 5, 6 represent the domino halves carrying the corresponding number of dots. The edges of the graph represent the dominoes (e.g. $\underset{}{0\_\!\_1}$ represents $\Box\bullet$, $\underset{}{2\_\!\_1}$ represents $\boxdot\!\therefore$ and

so on). Double dominoes are represented by loops (for example $\underset{5}{Q}$ represents $\boxdot\!\boxdot$).

Each edge carries an arrow pointing to the edge which represents the next domino in the chain (e.g. ▢⁞ ▢⁞ ▢⁞ ▢⁞ . . . will be represented by

●──▶●──▶●⟍⭕
0   1   2 ⟍5). This makes the graph oriented.

If C is a domino chain containing all pieces of the set, then in the corresponding graph G every pair of vertices is joined by an oriented edge, and at each vertex there is an oriented loop. Moreover G contains a so-called Eulerian path, that is: It is possible to draw, without lifting the pencil, a path P which passes along each edge and each loop of the graph exactly once.

Thus the problem of constructing the domino chain C is reduced to the problem of constructing an Eulerian path P for the graph G shown below:

Using Fig. 2.5(a) it is easy to find several Eulerian paths on G. One of them is P, consisting of the following steps:

$$0 \to 0, \ 0 \to 1, \ 1 \to 1, \ 1 \to 2, \ 2 \to 2, \ 2 \to 3, \ 3 \to 3,$$
$$3 \to 4, \ 4 \to 4, \ 4 \to 5, \ 5 \to 5, \ 5 \to 6, \ 6 \to 6, \ 6 \to 0,$$
$$0 \to 2, \ 2 \to 4, \ 4 \to 6, \ 6 \to 1, \ 1 \to 3, \ 3 \to 5, \ 5 \to 0,$$
$$0 \to 3, \ 3 \to 6, \ 6 \to 2, \ 2 \to 5, \ 5 \to 1, \ 1 \to 4, \ 4 \to 0,$$

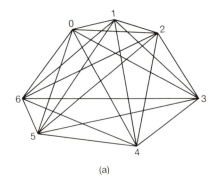

(a)

Fig. 2.5(a)

The domino chain C, obtained from P by replacing each oriented edge and loop by the corresponding domino piece, gives an affirmative answer to part (a) of Problem 51.

(b) A domino chain, constructed in the required manner from the pieces with 0 to 5 dots on their halves, would correspond to an Eulerian path P' on the graph G' (Fig. 2.5(b)).

Suppose that P' has been constructed. Let i be any vertex which is not on either end of P'.

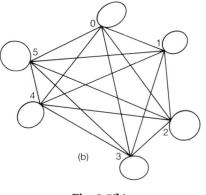

Fig. 2.5(b)

An oriented edge of P' leading *to i* is followed by an oriented edge leading *from i*. Hence the number of edges through i has to be even. However, there are five edges through each vertex of G'. This implies that P' cannot be drawn, and equivalently:

It is not possible to construct a domino chain from all pieces with 0 to 5 dots on their halves such that touching halves carry the same number of dots.

Problem 52 (E)

(a) The following geometric interpretation makes the solution easy:

For any two positive numbers x, y the number $\sqrt{(x^2 + y^2)}$ can be considered as the length of the hypotenuse in a right-angled triangle with the two other sides of length x and y. The numbers $\sqrt{(a_1^2 + b_1^2)}, \sqrt{(a_2^2 + b_2^2)}, \ldots$ $\sqrt{(a_n^2 + b_n^2)}$ are illustrated in Fig. 2.6.

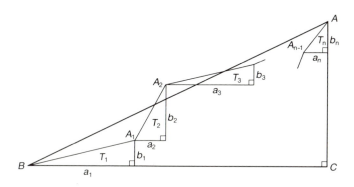

Fig. 2.6

In Fig. 2.6 the right-angled triangles T_i have sides a_i parallel to a_1 and b_i parallel to b_1 for $i = 2, 3, \ldots, n$. The hypotenuses of T_i form a polygonal line $p = BA_1A_2A_3 \cdots A_{n-1}A$. Thus

$$p = \sqrt{(a_1^2 + b_1^2)} + \sqrt{(a_2^2 + b_2^2)} + \sqrt{(a_3^2 + b_3^2)} + \cdots + \sqrt{(a_n^2 + b_n^2)}.$$

By producing the sides a_1 and b_n a right-angled triangle ABC is obtained. Its sides are:

$$BC = a_1 + a_2 + \cdots + a_n, \ CA = b_1 + b_2 + \cdots + b_n$$

and

$$BA = \sqrt{[(a_1 + a_2 + \cdots + a_n)^2 + (b_1 + b_2 + \cdots + b_n)^2]}.$$

Clearly, $BA \le p$, thus

$$\sqrt{[(a_1 + a_2 + \cdots + a_n)^2 + (b_1 + b_2 + \cdots + b_n)^2]} \le \sqrt{(a_1^2 + b_1^2)}$$
$$+ \sqrt{(a_2^2 + b_2^2)} + \cdots + \sqrt{(a_n^2 + b_n^2)}. \tag{4}$$

(b) Equality in (4) holds if and only if all vertices $A_1, A_2, \ldots, A_{n-1}$ belong to BA. In that case all triangles T_i are similar, hence their corresponding sides are proportional. In other words:

$$\sqrt{[(a_1 + a_2 + \cdots + a_n)^2 + (b_1 + b_2 + \cdots + b_n)^2]} = \sqrt{(a_1^2 + b_1^2)}$$
$$+ \sqrt{(a_2^2 + b_2^2)} + \cdots + \sqrt{(a_n^2 + b_n^2)}.$$

if and only if $a_1 : a_2 : \cdots : a_n = b_1 : b_2 : \cdots : b_n$.

Problem 53

For $n = 1$ the problem is trivial.

Consider the case $n = 2$. The equation $a_1^2 + a_2^2 = b^2$ can be rewritten as

$$\left(\frac{a_1}{b}\right)^2 + \left(\frac{a_2}{b}\right)^2 = 1,$$

or, by putting

$$\frac{a_1}{b} = x, \qquad \frac{a_2}{b} = y \tag{5}$$

as

$$x^2 + y^2 = 1. \tag{6}$$

(6) is the equation of the unit circle c with centre O $(0,0)$ in the two-dimensional Cartesian coordinate system. Hence Problem 53 in case $n = 2$ can be rephrased as follows.

Problem E

Prove that c contains infinitely many 'rational points', that is points with rational coordinates (x, y).

Solution to E

The rational point $A(-1, 0)$ lies on c. Any straight line through $A(-1, 0)$ with rational slope s cuts c in a rational point B, different from A (Fig. 2.7).

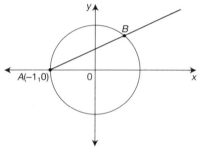

Fig. 2.7

To prove this, let s be any rational number, say $s = p/q$. The coordinates of B satisfy the system of equations

$$\left.\begin{array}{l} x^2 + y^2 = 1 \\[2mm] y = s(x + 1) \end{array}\right\} \tag{7}$$

The solutions of (7) are $x_1 = -1$, $y_1 = 0$ and

$$x_2 = \frac{1 - s^2}{1 + s^2}, \qquad y_2 = \frac{2s}{1 + s^2}.$$

x_2, y_2 are the coordinates of B. They can be expressed in terms of p and q.

$$x_2 = \frac{q^2 - p^2}{q^2 + p^2}, \qquad y_2 = \frac{2pq}{q^2 + p^2}. \tag{8}$$

Since there are infinitely many rational numbers p/q, the circle c contains infinitely many rational points $B(x_2, y_2)$. This solves Problem E.

By comparing (8) and (5) it follows that

$$a_1 = q^2 - p^2, \qquad a_2 = 2pq \qquad \text{and} \qquad b = q^2 + p^2$$

are integers satisfying the original equation $a_1^2 + a_2^2 = b^2$. If $q > p > 0$ then a_1, a_2 and b are positive. There are infinitely many pairs p, q satisfying these conditions, hence the equation has infinitely many solutions in positive integers. Thus Problem 53 is solved for $n = 2$.

The above method can be generalized for $n \geq 3$. In the general case (7) is replaced by the system of simultaneous equations

$$x_1^2 + x_2^2 + \cdots x_n^2 = 1$$

$$x_2 = s_2(x_1 + 1)$$

$$x_3 = s_3(x_1 + 1)$$

$$\vdots$$

$$x_n = s_n(x_1 + 1),$$

where $x_i = a_i/b$ for $i = 1, \ldots, n$.

Following the same arguments as in the case $n = 2$ one deduces that the equation

$$a_1^2 + a_2^2 + \cdots + a_n^2 = b^2$$

has infinitely many solutions in positive integers.

Problem 54

Construct a graph G such that:

- the vertices of G represent the inhabitants of the estate, and
- two vertices are connected by an edge if and only if the corresponding inhabitants are acquainted.

A sequence of distinct edges connecting a pair of vertices is called a path joining the vertices. Since any item of news is eventually transmitted to all inhabitants, any two vertices of G are connected by a path. A graph with this property is called connected. The number of edges in a path is the length of the path. The length of the shortest path connecting two vertices A, B is the distance between A and B.

Problem 54 can be restated as the following problem on graphs.

Problem F

Let G be a connected graph with 1000 vertices. Prove that there is a set S of 90 vertices in G satisfying the following property:
(P) For any vertex A in G there is at least one vertex B in S whose distance from A is at most 10.

Notice that G may contain paths whose end-points coincide; such paths are called cycles (Fig. 2.8(a)).
Otherwise G is a graph without cycles. Graphs without cycles are called trees (Fig. 2.8(b)).

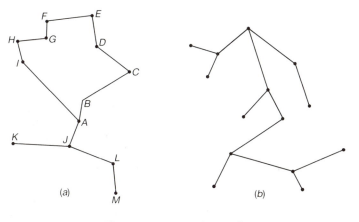

Fig. 2.8(a)   Fig. 2.8(b)

It is sufficient to prove the existence of the set S in the case when G is a tree.
For, suppose that G contains some cycles. By removing from G an edge contained in one of its cycles a new graph G_1 is constructed. G_1 is still connected but contains a smaller number of cycles than G. By repeating the same procedure, after a finite number of steps G is transformed into a tree G_k. The graphs G and G_k have the same vertices and for any two vertices their mutual distance in G does not exceed their mutual distance in G_k. Hence if S is a set of 90 vertices in G_k satisfying (P), then S, considered as a set of vertices in G, has the same property (P).
Let G be a tree with 1000 vertices, and let X and Y be two vertices at the greatest possible distance d in G. If $d \leq 10$, then the existence of S is trivial. So, suppose that $d > 10$. Denote the vertices on the shortest path from X to Y by $X, A_1, A_2, \ldots, A_{10}, A_{11}, \ldots, Y$.
Sever the edge $A_{10}A_{11}$, dividing G into two trees: G^{I}, containing A_{10}, and G^{II}.

In G^I the distance of any vertex M from A_{10} is less than or equal to 10; otherwise the distance of M from Y would be greater than d, the greatest possible distance in G. Choose A_{10} as the first vertex of S.

G^I contains at least 11 vertices, hence the number of vertices in G^{II} is at most $1000 - 11 = 989$. Repeat the same procedure for G^{II}; as a result G^{II} is divided into two trees G^{III} and G^{IV}, such that G^{III} has at least 11 vertices, one of which is at a distance not greater than 10 from all vertices in G^{III}. Choose this point as the second vertex of S. G^{IV} contains at most $1000 - 2 \cdot 11 = 978$ vertices.

After 89 similar steps G is divided into 90 trees. Each of the first 89 trees contains one vertex of S. The 90th tree G^* has at most $1000 - 89 \times 11 = 21$ vertices. Let X^* and Y^* be vertices at maximal distance d^* in G^* and, if $d^* > 10$, let A_{90} be the 11th vertex on the shortest path from X^* to Y^* (including X^*). Take A_{90} as the 90th vertex of S. If $d^* \le 10$, take any vertex of G^* as A_{90}.

Obviously, S satisfies (P).

Remark: The solution to Problem 54 is the best possible: In the general case it is not possible to construct a set S of less than 90 elements satisfying condition (P). Figure 9 illustrates this point. It shows a tree T with 991 vertices on 90 branches, emerging from a common vertex N. Each branch carries 12 vertices, including N. It is impossible to construct a set S' of 89 vertices in T such that any vertex of T is at a distance of at most 10 from some vertex in S.

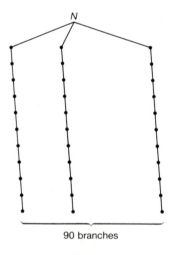

90 branches

Fig. 2.9

Problem 55

(a) Let the length of the line segment AB be 1. Denote the lengths of the parts into which AB is divided by x_1, x_2, and x_3. The three parts of AB form a triangle if and only if

$$x_1 < x_2 + x_3, x_2 < x_1 + x_3 \text{ and } x_3 < x_1 + x_2,$$

or, equivalently,

$$x_1 < \tfrac{1}{2}, x_2 < \tfrac{1}{2} \text{ and } x_3 < \tfrac{1}{2}. \tag{9}$$

Since x_2, x_2 and x_3 are three variables with a constant sum, they can be considered as the barycentric coordinates of a point P in the plane.

The barycentric coordinate system in the plane, invented by Möbius (nineteenth century), consists of three coordinate axes x_1, x_2, x_3 forming an equilateral triangle $A_1 A_2 A_3$. The coordinates x_1, x_2, x_3 of any point P in the plane are the distances of P from the axes respectively ($x_i < 0$ if and only if P and A_i are on different sides of axis x_i).

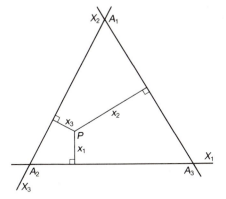

Fig. 2.10

Let b and h be, respectively, the base and height of $\triangle A_1 A_2 A_3$. From

$$\tfrac{1}{2}bh = \text{Area } A_1 A_2 A_3 = \text{Area } A_2 P A_3 + \text{Area } A_3 P A_1 + \text{Area } A_1 P A_2$$

$$= \tfrac{1}{2}bx_1 + \tfrac{1}{2}bx_2 + \tfrac{1}{2}bx_3,$$

it follows that

$$x_1 + x_2 + x_3 = h$$

for any point $P(x_1, x_2, x_3)$.

In the barycentric coordinate system corresponding to triangle $A_1 A_2 A_3$ with altitude $h = 1$, conditions (9) are satisfied if and only if P is in the region bounded by straight lines with equations $x_i = \frac{1}{2}$ for $i = 1, 2, 3$ (Fig. 2.11). Thus Problem 55(a) is transformed into a new problem.

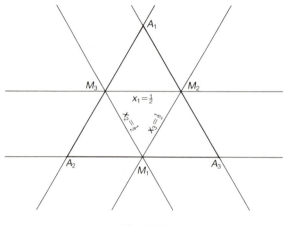

Fig. 2.11

Problem G

If P is a point inside triangle $A_1 A_2 A_3$, find the probability of P lying inside triangle $M_1 M_2 M_3$ formed by the midpoints of the sides $A_2 A_3$, $A_3 A_1$ and $A_1 A_2$.

The answer to the above question is easily obtained: denote the area of $\triangle A_1 A_2 A_3$ by $\mathbb{A}$. Then the area of $\triangle M_1 M_2 M_3$ is $\frac{1}{4}\mathbb{A}$, and the probability of P lying in $\triangle M_1 M_2 M_3$ is

$$p_3 = \tfrac{1}{4}.$$

This solves Problem G as well as Problem 55(a).

(b) Problem 55(a) can be generalized as follows:

$n - 1$ points are chosen at random on a straight line segment, dividing it into n points. What is the probability p_n that an n-gon can be constructed from the n parts as sides?

If $n = 4$, the parts x_1, x_2, x_3 and x_4 of the line segment AB (of length 1) can be considered as the barycentric coordinates corresponding to a regular tetrahedron $T = A_1A_2A_3A_4$ of height 1. In this coordinate system x_i is the distance of a point from the face of T, opposite A_i (Fig. 2.12).

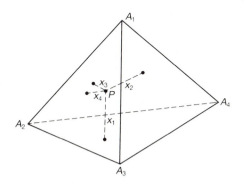

Fig. 2.12

The condition that the four parts form a quadrilateral is equivalent to

$$x_1 < x_2 + x_3 + x_4, \quad x_2 < x_1 + x_3 + x_4,$$
$$x_3 < x_1 + x_2 + x_4, \quad x_4 < x_1 + x_2 + x_3,$$

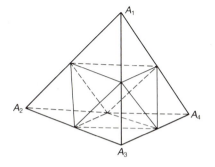

Fig. 2.13

that is, to

$$x_i < \tfrac{1}{2} \qquad \text{for } i = 1, 2, 3, 4. \tag{10}$$

The region R defined by (10) is obtained from T by removing at each vertex a tetrahedron T_i with edges equal to half of the edges of T (Fig. 2.13).

Thus the probability of conditions (10) being satisfied is

$$p_4 = \frac{\text{Volume } T - 4 \times \text{volume } T_i}{\text{Volume } T} = 1 - 4(\tfrac{1}{2})^3.$$

In the general case, a barycentric coordinate system can be set up in n-dimensional space. The rôle of the tetrahedra T and T_i is taken over by n-dimensional simplexes S and S_i. By similar arguments one deduces that

$$p_n = \frac{\text{Volume } S - n \times \text{volume } S_i}{\text{Volume } S} = 1 - n(\tfrac{1}{2})^{n-1}.$$

Section 2: Extending the field of investigation

Problem 56

To answer the question by considering prime numbers only seems difficult because not much is known about the distribution of prime numbers on the number line. However, the sequence of all prime numbers

$$2, 3, 5, 7, 11, 13, \ldots$$

has a property shared by some other sequences: The difference between any two consecutive terms, starting from the second, is at least 2.

The above property suggests the investigation of the following, more general problem:

Problem H

Let $a_1, a_2, a_3, \ldots, a_n, a_{n+1}, \ldots$ be a sequence of natural numbers such that

$$a_1 = 2, a_2 = 3 \text{ and } a_{n+1} - a_n \geq 2 \text{ for } n = 2, 3, \ldots. \tag{11}$$

Denote the sum $a_1 + a_2 + \cdots + a_n$ by σ_n for $n = 1, 2, \ldots$. Is it true that there is a square number between σ_n and σ_{n+1} for any natural number n?

The solution of problem H will be carried out in several steps.

Step 1:   Suppose that there exists a natural number n such that there is no square number between σ_n and σ_{n+1}. In that case, for some natural number k the following relations hold:

$$k^2 \leq \sigma_n < \sigma_{n+1} \leq (k + 1)^2. \tag{12}$$

This implies that

$$\underbrace{\sigma_{n+1} - \sigma_n}_{a_{n+1}} \leq (k + 1)^2 - k^2 = 2k + 1,$$
$$\phantom{\sigma_{n+1} - \sigma_n} \leq 2k + 1 \tag{13}$$

According to (11), $a_n \leq a_{n+1} - 2$ for $n \geq 2$. This, combined with (13), leads to the inequalities

$$\left.\begin{array}{c} a_{n+1} \leq 2k + 1 \\ a_n \leq 2k - 1 \\ a_{n-1} \leq 2k - 3 \\ \vdots \\ a_2 \leq 2k + 1 - 2(n - 1) = x. \end{array}\right\} \tag{14}$$

Two cases will be distinguished: (i) $x = 3$ and (ii) $x > 3$.

Step 2:   Consider case (i): $x = 3$.
   This implies that $a_2 = 3$. Hence equality must hold in all relations in (14). Thus

$$a_{n+1} + a_n + \cdots + a_2 + a_1 = (2k + 1) + (2k - 1) + \cdots + 3 + 2$$
$$> (2k +) + (2k - 1) + \cdots + 3 + 1$$
$$= (k + 1)^2.$$

That is

$$\sigma_{n+1} > (k + 1)^2,$$

contradicting (12).
   Case (i) cannot occur.

Step 3:   Consider case (ii): $x > 3$.
    In this case $x \geq 5$, and

$$a_n + a_{n-1} + \cdots + a_2 + a_1 < (2k - 1) + (2k - 3) + \cdots + 5 + 3 + 1 = k^2.$$

Thus

$$\sigma_n < k^2,$$

contradicting (12).
    Case (ii) cannot occur either.
    The above investigations show that for any natural number n there is a square number between σ_n and σ_{n+1}. This solves Problem H, and also Problem 56, as a special case of Problem H.

Problem 57

Extend the investigations to three-dimensional space: Consider a regular dodecahedron D. The circles inscribed in the faces of D form a set S' such that each circle of S' touches exactly five circles of the set. The sphere σ, touching all edges of D, carries the circles of S'. Let AB be a diameter of σ passing through the centres of two opposite faces of D. Denote by π the plane touching σ at B.

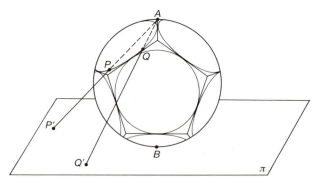

Fig. 2.14

From A as centre construct the stereographic projections of the circles of S' onto π. (The stereographic projection P of a point P' on σ is the intersection of the straight line AP' with π.) Since the circles of S' do not pass through A, their projections are circles on π (see Appendix I). The set S of these circles has the required property: each of them touches five circles, all in different points.

Problem 58

Together with $a_n = (2 + \sqrt{3})^n$ consider the numbers $b_n = (2 - \sqrt{3})^n$. Since

$$a_n = \binom{n}{0} 2^n (\sqrt{3})^0 + \binom{n}{1} 2^{n-1} (\sqrt{3})^1 + \cdots + \binom{n}{n} 2^0 (\sqrt{3})^n,$$

and

$$b_n = \binom{n}{0} 2^n (\sqrt{3})^0 - \binom{n}{1} 2^{n-1} (\sqrt{3})^1 + \cdots + (-1)^n \binom{n}{n} 2^0 (\sqrt{3})^n,$$

the sum

$$a_n + b_n = 2 \left(\binom{n}{0} 2^n + \binom{n}{2} 2^{n-1} \cdot 3 + \cdots + \binom{n}{2k} 2^{n-2k} \cdot 3^k + \cdots \right)$$

$$= 2t.$$

Thus $a_n + b_n$ is an even positive number. $2 - \sqrt{3}$ is a positive number, less than 1. Hence $0 < b_n < 1$ and $[a_n] = [2t - b_n] = 2t - 1$.

$[a_n]$ is an odd natural number for all $n = 1, 2, 3, \ldots$.

Problem 59

The theorem of Desargues in the plane can be proved, for example, by introducing coordinates. However, coordinates can be avoided if investigations are extended to three-dimensional space:

Let π be the plane containing triangles ABC and $A'B'C'$. Draw a straight line ℓ through S_1 outside π, and mark two points S and S' on ℓ such that S is between S_1 and S'. The points S, S', A and A' belong to a common plane $(S'S_1A)$ and S and A are separated by the straight line $S'A'$ (Fig. 2.15). Therefore,

$$SA \text{ and } S'A' \text{ meet in a point } A_1.$$

Similarly,

$$SB \text{ and } S'B' \text{ meet in a point } B_1$$

and

$$SC \text{ and } S'C' \text{ meet in a point } C_1.$$

This points A_1, B_1, and P are common to the planes SAB and $S'A'B'$.

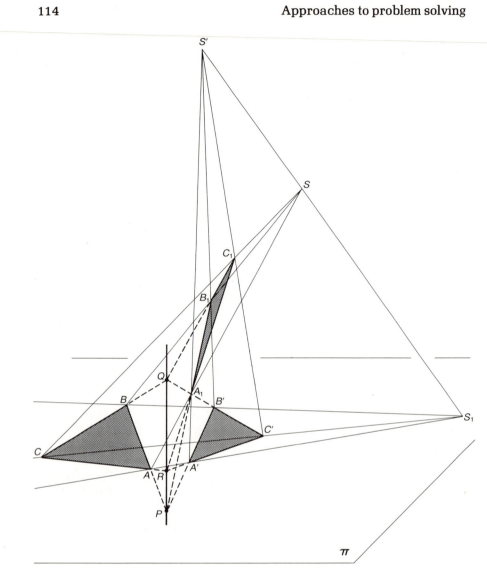

Fig. 2.15

Therefore A_1, B_1, and P are collinear. Thus A_1B_1 meets π in P, that is, in the point where AB and $A'B'$ intersect.

For similar reasons,

$$B_1C_1 \text{ meets } \pi \text{ in } Q, \text{ common to } BC \text{ and } B'C'$$

and

$$C_1A_1 \text{ meets } \pi \text{ in } R, \text{ common to } CA \text{ and } C'A'.$$

Thus P, Q and R belong to both planes $A_1B_1C_1$ and π. Since the intersection of two non-parallel planes is a straight line, this implies that P, Q, and R are collinear.

Remark: The theorem of Desargues for two triangles in different planes is treated in Problem 48.

Section 3: The use of invariants of transformations

Problem 60

Suppose that after k encounters of pairs of chameleons with different colours, there are y_k yellow, g_k grey and b_k blue chameleons on the island. At the $(k + 1)$st encounter the triple $T_k = (y_k, g_k, b_k)$ will change into one of the following triples: (y_k-1, g_k-1, b_k+2), (y_k-1, g_k+2, b_k-1) or (y_k+2, g_k-1, b_k-1).

Note that the difference $y_{k+1} - g_{k+1}$ in the corresponding cases is equal to $y_k - g_k, y_k - g_k - 3, y_k - g_k + 3$ respectively. In other words, the remainder of $y_k - g_k$ after division by 3 remains invariant under the transformation $(y_k, g_k, b_k) \rightarrow (y_{k+1}, g_{k+1}, b_{k+1})$.

At the start $y_0 - g_0 = 17 - 15 = 2$. Therefore

$$y_k - g_k \equiv 2 \text{ (modulo 3)}$$

for all $k = 0, 1, 2, \ldots$.

If after t encounters all chameleons on the island had the same colour, then the difference $y_t - g_t$ would be equal to one of the numbers, 0, -45 or 45. Since none of these numbers leaves the remainder 2 after division by 3, it follows that at no time can all chameleons on the island be of the same colour.

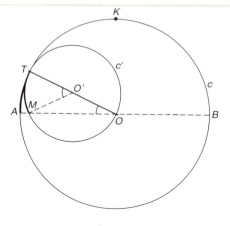

Fig. 2.16

Problem 61 (E)

Suppose that initially M is on c, at a point A. At an arbitrary instant, denote the point of contact between c' and c by T, and the centre of c' by O'.

Figure 2.16 shows a position of T on its way from A to K, where $\sphericalangle AOK = 90°$.

While T is moving along $\overset{\frown}{AK}$, the transformation carrying M into new positions changes the angles $\sphericalangle MO'T$ and $\sphericalangle AOT$ according to the rule:

$$\sphericalangle MO'T = \frac{\overset{\frown}{MT}}{r'} \text{ and } \sphericalangle AOT = \frac{\overset{\frown}{AT}}{r}.$$

By assumption $r = 2r'$. Since r' rolls without slipping, $\overset{\frown}{MT} = \overset{\frown}{AT}$. Thus

$$\sphericalangle MO'T = 2\sphericalangle AOT, \qquad (15)$$

that is, the ratio $\sphericalangle MO'T : \sphericalangle AOT$ remains invariant.

The point O is on c'; therefore

$$\sphericalangle MO'T = 2\sphericalangle MOT. \qquad (16)$$

(15) and (16) imply that $\sphericalangle TOM = \sphericalangle TOA$, that is, M lies on the radius AO of C.

It is now easy to determine the path of M while T traverses the complete circumference c:

- when T moves along $\overset{\frown}{AK}$, the point M describes the radius AO of c;

- when T proceeds from K to the point B on c, diametrically opposite to A, the point M moves from O to B;
- when T describes the remaining half of c, M returns along BA to its original position.

It is interesting to note that M performs a linear motion.

Section 4: The use of extremal elements

Problem 62

Suppose that the equation

$$x^2 + y^2 = 3(z^2 + u^2) \qquad (17)$$

has solutions in positive integers. Among them must be one with the smallest possible value of x. Denote this solution by (x_1, y_1, z_1, u_1). From

$$x_1^2 + y_1^2 = 3(z_1^2 + u_1^2) \qquad (18)$$

it follows that the sum $x_1^2 + y_1^2$ is divisible by 3. This can be true only if both x_1 and y_1 are divisible by 3. (If an integer is not divisible by 3, then its square is of the form $3k + 1$ for some integer k.)

Thus, $x_1 = 3x_2$ and $y_1 = 3y_2$ for some positive integers x_2 and y_2. By substituting these expressions into (18) we get

$$9x_2^2 + 9y_2^2 = 3(z_1^2 + u_1^2),$$

or

$$3x_2^2 + 3y_2^2 = z_1^2 + u_1^2. \qquad (19)$$

(19) implies that z_1 and u_1 are divisible by 3. Hence $z_1 = 3z_2$ and $u_1 = 3u_2$ for some positive integers z_2 and u_2. Substituting these into (19) leads to

$$x_2^2 + y_2^2 = 3(z_2^2 + u_2^2).$$

In other words (x_2, y_2, z_2, u_2) is a solution of (17). In this solution $x_2 = \frac{1}{3}x_1 < x_1$, contradicting the assumption that x_1 was the smallest possible value for all solutions of (17).

Hence, equation (17) has no solution in positive integers.

Problem 63 (E)

The n points form finitely many triangles; therefore among them there is one with the largest area. Denote this triangle by $T = MNL$. Through each vertex

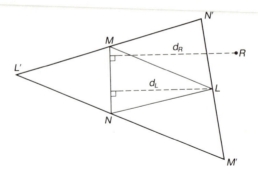

Fig. 2.17

of T draw a parallel to the opposite side of T. In this way a triangle $T' = M'N'L'$ is constructed with M, N, L representing the midpoints of the sides $N'L'$, $L'M'$ and $M'N'$ respectively (Fig. 2.17).

The area of T' is four times the area of T, hence it is not greater than 4. Our aim is to show that no point of P is outside T'.

Suppose that there exists a point R of P outside T'. In this case at least one vertex of T' is separated from R by the opposite side of T' (i.e. the line segment, joining this vertex to R, intersects the opposite side of T'). Without loss of generality let us suppose that $M'N'$ separates R from L'. This implies that the distance d_R of MN from R is greater than its distance d_L from L. The distances d_R and d_L are the altitudes of the triangles RMN and LMN, corresponding to their common side MN. Thus

$$d_R > d_L,$$

which implies that

$$\text{Area } RMN = \tfrac{1}{2}MNd_R > \tfrac{1}{2}MNd_L = \text{Area } LMN.$$

The last inequality leads to a contradiction, since T is the triangle of largest area with vertices in P. We have proved that R cannot lie outside T'.

Section 5: The method of infinite descent

Problem 64 (E)

Suppose that a given cube $C = KLMNOPQR$ can be divided into finitely many smaller cubes C_i, $i = 1, 2, \ldots, n$, all of different size. In that case some of the cubes C_i have faces on the face $KLMN$ of C, and divide $KLMN$ into finitely many squares S_j, $j = 1, 2, \ldots, k$, all of different size. Denote

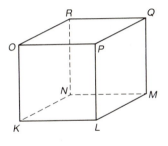

Fig. 2.18

the smallest of these squares by S_m. Our first aim is to show that S_m lies inside *KLMN*.

Suppose that a side $K_m L_m$ of S_m belongs to a side of *KLMN*, say *KL*. In that case the opposite side $M_m N_m$ of S_m is enclosed in a 'hole' between two of the squares S_j larger than S_m (Fig. 2.19). Hence $M_m N_m$ carries the sides of at least two squares S_j, say $S_{j'}$ and $S_{j''}$. Thus $S_{j'}$ and $S_{j''}$ are smaller than S_m, which is a contradiction.

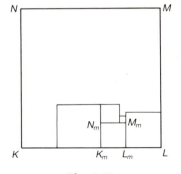

Fig. 2.19

The above contradiction shows that $S_m = K_m L_m M_m N_m$ lies inside *KLMN*. Let $C_m = K_m L_m M_m N_m O_m P_m Q_m R_m$ be the cube with face S_m. The square S_m is surrounded by larger squares $S_{j'}$. The cubes with faces $S_{j'}$ are all larger than C_m; hence the face $O_m P_m Q_m R_m$ of C_m lies in a hole between some of the cubes C_i. This implies that $O_m P_m Q_m R_m$ carries a finite number of cubes C_i, all smaller than C_m, dividing it into squares of different size. Among them there is a smallest square; this square belongs to a cube $C_{m'}$ smaller than C_m. By repeating the same argument we can construct an infinite sequence of cubes:

$$C_m > C_{m'} > C_{m''} > \cdots .$$

However, according to our assumption, C was divided into a finite number of cubes. This contradiction shows that a cube cannot be divided into finitely many cubes all of different size.

Problem 65 (E)

(a) Suppose that we have constructed in a two-dimensional Cartesian coordinate system a regular pentagon $P_1 = A_1B_1C_1D_1E_1$ whose vertices are lattice points.

Draw the vectors:
$A_1A_2 = B_1C_1$, $B_1B_2 = C_1D_1$, $C_1C_2 = D_1E_1$, $D_1D_2 = E_1A_1$ and $E_1E_2 = A_1B_1$. A_2, B_2, C_2, D_2 and E_2 are also lattice points, and they are the vertices of a regular pentagon P_2. (Why?)

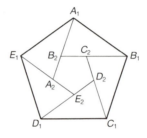

Fig. 2.20

Similarly, using P_2, we can construct a regular pentagon P_3 whose vertices are lattice points. By iterating this procedure we obtain an infinite sequence of regular pentagons

$$P_1, P_2, P_3, \ldots, P_n, \ldots$$

whose vertices are lattice points. Denote by a_n the side length of P_n for $n = 1$, $2, \ldots$. The squares $a_1^2, a_2^2, a_3^2, \ldots, a_n^2, \ldots$ form an infinite decreasing sequence of natural numbers. However, any decreasing sequence of natural numbers has a smallest element; hence we have reached a contradiction.

Thus P_1 cannot be constructed. The answer to part (a) of Problem 65 is 'No'.

(b) By the same method as in (a) it can be proved that:

1. For any $n > 6$ there is no regular n-gon whose vertices are lattice points in a two-dimensional Cartesian coordinate system.

The above method does not apply to n-gons with $n = 3$ or 6. However, it is not difficult to obtain a similar result:

2.   There is no regular triangle, or hexagon whose vertices are lattice points in a two-dimensional coordinate system.

To prove (2) for $n = 3$ suppose that there is an equilateral triangle T whose vertices are lattice points. In that case the square of T's side length, a^2, is an integer, and the area of T is an irrational number.

On the other hand, by circumscribing a rectangle R about T, it is easy to deduce that the area of T must be a rational number.

This contradiction shows that T does not exist.

Let n be 6. In a regular hexagon $ABCDEF$ the vertices A, C and E form an equilateral triangle. Since there is no equilateral triangle whose vertices are lattice points, the same applies to a regular hexagon.

3.   For $n = 4$ there are regular n-gons, that is squares whose vertices are lattice points.

Thus the general result is the following:

The only regular polygons whose vertices are lattice points in a two-dimensional Cartesian coordinate system are the squares.

Section 6: Mathematical induction

Problem 65

The statement will be proved by induction on k:

$$\text{For } k = 1, S_{1,n} = 1 + 2 + \cdots + n = \frac{n(n + 1)}{2} = \tfrac{1}{2}n^2 + \tfrac{1}{2}n.$$

$S_{1,n}$ is a polynomial in n, of degree 2, with leading coefficient $\tfrac{1}{2}$. Thus the statement is correct if $k = 1$.

Suppose that the statement is true for any k less than t. Our aim is to deduce that the statement holds for $k = t$.

Recall that

$$(a + 1)^m = \binom{m}{0} a^m + \binom{m}{1} a^{m-1} + \cdots + \binom{m}{m} a^0. \qquad (20)$$

By writing down equation (20) for $m = t + 1$ and $a = n, n-1, \ldots, 2$, and by adding these equations we get

$$(n+1)^{t+1} = n^{t+1} + \binom{t+1}{1} n^t + \cdots + \binom{t+1}{t+1} n^0$$

$$n^{t+1} = (n-1)^{t+1} + \binom{t+1}{1}(n-1)^t + \cdots + \binom{t+1}{t+1}(n-1)^0$$

$$(n-1)^{t+1} = (n-2)^{t+1} + \binom{t+1}{1}(n-2)^t + \cdots + \binom{t+1}{t+1}(n-2)^0$$

$$\vdots$$

$$2^{t+1} = 1^{t+1} + \binom{t+1}{1}1^t + \cdots + \binom{t+1}{t+1}1^0$$

$$(n+1)^{t+1} = 1 + \binom{t+1}{1}S_{t,n} + \binom{t+1}{2}S_{t-1,n} + \cdots + \binom{t+1}{t+1}S_{0,n}$$

Thus

$$S_{t,n} = \frac{1}{t+1}\left[(n+1)^{t+1} - \binom{t+1}{2}S_{t-1,n} - \binom{t+1}{3}S_{t-2,n} - \cdots\right.$$
$$\left. - \binom{t+1}{t+1}S_{0,n} - 1\right]. \tag{21}$$

According to the induction hypothesis, the sums $S_{t-1,n}$, $S_{t-2,n}$, . . ., $S_{0,n}$ are polynomials in n of degrees t, $t-1$, . . ., 1, respectively.

$$(n+1)^{t+1} = n^{t+1} + \binom{t+1}{1}n^t + \cdots + \binom{t+1}{t+1}n^0.$$

These facts and (21) imply that $S_{t,n}$ is a polynomial in n of degree $t+1$, with leading coefficient $1/(t+1)$.

Problem 67

We shall use induction on the dimension of the space. In other words, the solution of Problem 67 will be deduced from the solution of the analogous problem in the plane (that is, in two-dimensional space). The latter problem will be proved by using the solution of the corresponding problem on a straight line (i.e. one-dimensional space). The machinery set up to prove problem P_1 will extend to prove problems P_2 and P_3.

(a) The one-dimensional problem P_1 The endpoints of a straight line segment S are numbered 1 and 2. S is divided into a finite number of smaller segments S_i, and the endpoints of S_i are numbered 1 or 2 in an arbitrary

manner. Prove that there is at least one segment S_i with endpoints carrying different numbers.

Proof of P_1. A division point A_j of S is said to be incident with a segment S_i if A_j is an endpoint of S_i. In this case the pair (A_j, S_i) is called an incident point–segment pair. Denote (A_j, S_i) by $(1, S_i)$ if A_j is numbered 1 and by $(2, S_i)$ if A_j is numbered 2. We shall count the number n of incident pairs $(1, S_i)$. This can be done in two different ways:

1. If n_1 denotes the number of segments S_i with one endpoint 1, and n_2 the number of segments with two endpoints 1, then

$$n = n_1 \cdot 1 + n_2 \cdot 2. \tag{22}$$

2. All points 1, apart from the endpoint of S labelled 1, belong to two segments S_i. Thus, if n^* is the number of 1s on S, then

$$n = (n^* - 1) \cdot 2 + 1. \tag{23}$$

By combining (22) and (23) we find that

$$n_1 = 2(n^* - n_2) - 1. \tag{24}$$

Thus n_1 is an odd number; hence $n_1 \geq 1$. In other words, there is at least one small segment S_i with endpoints labelled 1 and 2.

(b) *The two-dimensional problem P_2* The vertices of a triangle t are labelled 1, 2 and 3. Triangle t is divided into a finite number of smaller triangles t_i such that two triangles t_i have either no point in common, or only a vertex in common, or only a side in common. The vertices of t_i are labelled 1, 2 or 3 subject to the following restriction: vertices of t_i belonging to a side of t with endpoints x, y must carry one of the labels x, y.

Prove that there is at least one small triangle t_i whose vertices carry different numbers.

Proof of P_2. A line segment $\overline{A_j B_k}$ and a triangle t_i will be called incident if $\overline{A_j B_k}$ is a side of t_i. The number m of incident pairs $(\overline{12}, t_i)$, where 1 and 2 are the labels of A_j and B_k respectively, can be counted in two different ways:

1. If m_1 is the number of triangles t_i with one side $\overline{12}$, and m_2 the number of t_i with two sides $\overline{12}$, then

$$m = m_1 \cdot 1 + m_2 \cdot 2. \tag{25}$$

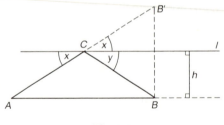

Fig. 2.21

Draw the line segment AB and a line ℓ parallel to AB at distance h from AB (Fig. 2.21). Our aim is to find on ℓ a point C such that the sum $AC + CB$ is minimal.

To find C it is useful to recall the following optical principle discovered by Heron of Alexandria (second century):

> A light-ray r, which emerges from A and reaches B after being reflected from ℓ, travels along the shortest path.

This implies that the point in which r meets ℓ is the vertex C of our triangle ABC. Having noted that, C can be constructed by applying another well-known property of reflected light rays:

The angle of incidence is equal to the angle of reflection.

Hence the angle x between ℓ and CA must be equal to the angle y between CB and ℓ. The construction of C is carried out as follows: Reflect B in ℓ and join the reflected point B' to A (Fig. 2.21). The intersection of $B'A$ with ℓ is C. The triangle ABC is isosceles.

It remains to be proved that for any point C^*, different from C, the following inequality holds:

$$AC^* + C^*B > AC + CB. \tag{30}$$

To prove (30), notice that C^*B and CB are equal to their mirror images C^*B' and CB' respectively. Thus $AC^* + C^*B = AC^* + C^*B'$ and $AC + CB = AC + CB' = AB'$.

A, C^* and B' are the vertices of triangle AC^*B'. In a triangle the sum of two sides is larger than the third. Thus

$$AC^* + C^*B' > AB'$$

or

$$AC^* + C^*B > AC + CB.$$

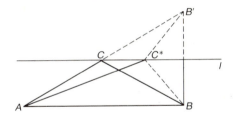

Fig. 2.22

We have proved that among all triangles of given base and area the isosceles triangle has the smallest perimeter.

(b) We shall describe two methods for solution.

Solution 1, extending the idea of (a), is by Fr Gabriel-Marie, the author of the book *Exercises de Géométrie*. Fix a point A' on the side BC of the acute-angled triangle ABC. Reflect A' in AB and AC to give A_1 and A_2 respectively.

If B'', C'' are any points on AB, respectively AC, then the perimeter of the triangle $A'B''C''$ is equal to $p_{A'} = A_1B'' + B''C'' + C''A_2$. $p_{A'}$ is the least when the points A_1, B'', C'' and A_2 are collinear. Denote the corresponding positions of B'' and C'' by B' and C'.

Triangle A_1AA_2 is isosceles with

$$A_1A = A_2A = A'A \quad \text{and} \quad \sphericalangle A_1AA_2 = 2\sphericalangle BAA' + 2\sphericalangle A'AC = 2\sphericalangle BAC.$$

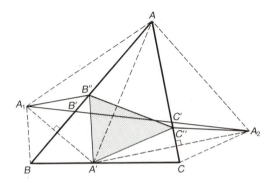

Fig. 2.23

Thus:

The perimeter of $A'B'C'$ is equal to $A_1A_2 = 2AA' \sin \angle BAC$. This implies that to obtain the overall minimum perimeter we must minimize AA'. The length of AA' will be the least when A' is the foot of the perpendicular from A to BC. Similarly, B' and C' must be the feet of the perpendiculars from B and C to the opposite sides of $\triangle ABC$. We have proved that the triangle of smallest perimeter inscribed in a given acute-angled triangle ABC is $\triangle A'B'C'$, formed by the feet A', B', C' of the altitudes AA', BB' and CC' of $\triangle ABC$.

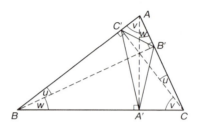

Fig. 2.24

Solution 2. Suppose that $A'B'C'$ is a triangle with smallest perimeter inscribed in a given acute-angled triangle ABC (Fig. 2.24). Heron's principle about the shortest path of reflected light rays suggests that the perimeter of $A'B'C'$ may be the path of a light ray reflected from the sides of ABC. In this case at each of the points A', B', C' the angle of incidence would be equal to the corresponding angle of reflection. This would imply the equality of the following pairs of angles:

$$\left.\begin{aligned} \angle BA'C' &= \angle B'A'C \\ \angle CB'A' &= \angle C'B'A \\ \angle AC'B' &= \angle A'C'B. \end{aligned}\right\} \tag{31}$$

Our aim is:

1. to find out whether there is a triangle $A'B'C'$ inscribed in ABC satisfying conditions (31), and, if so,
2. to investigate whether $A'B'C'$ has the smallest perimeter among all triangles inscribed in ABC.

1. Denote the angles of triangle ABC at A, B, and C by α, β, and γ respectively. Suppose that $A'B'C'$ is a triangle inscribed in ABC, satisfying

conditions (31). Put $\angle B'A'C = x$. The remaining angles in (31) can be expressed in terms of x, α, β and γ as follows:

$$\left.\begin{array}{l} \angle CB'A' = 180° - x - \gamma = \angle C'B'A \\ \angle AC'B' = x + \gamma - \alpha = \angle A'C'B \\ \angle BA'C' = 2a - x. \end{array}\right\} \tag{32}$$

Since $\angle BA'C' = \angle B'A'C = x$, it follows that $2\alpha - x = x$, that is $x = \alpha$. Thus the angles between the sides of $\triangle ABC$ and $\triangle A'B'C'$ are:

$$\left.\begin{array}{l} \angle BA'C' = \angle B'A'C = \alpha \\ \angle CB'A' = \angle C'B'A = \beta \\ \angle AC'B' = \angle A'C'B = \gamma \end{array}\right\} \tag{33}$$

The next step is to determine the position of the points A', B', C' on the sides of $\triangle ABC$.

Equations (33) imply that in the quadrilaterals $BCB'C'$, $ACA'C'$ and $ABA'B'$ opposite angles add up to $180°$. Hence these quadrilaterals are cyclic, and

$$\angle C'BB' = \angle C'CB' = u,$$
$$\angle C'AA' = \angle C'CA' = v,$$
$$\angle B'BA' = \angle A'AB' = w.$$

The sum $2u + 2v + 2w$ equals $180°$. Hence $u + v + w = 90°$. But $u + v = \gamma$, and therefore

$$\gamma + w = 90°.$$

Thus in $\triangle AA'C$ the angle $\angle AA'C$ is $90°$. In other words, AA' is the altitude of $\triangle ABC$ perpendicular to BC.

For similar reasons, BB' and CC' are the altitudes of triangle ABC perpendicular to AC and AB respectively. We have proved that there is exactly one triangle $A'B'C'$ satisfying conditions (31); its vertices are the feet of the altitudes of triangle ABC.

2. $A'B'C'$ is the triangle of the least possible perimeter inscribed in the acute-angled triangle ABC.

The following ingenious proof of the above statement is due to Schwarz (1843–1921):

Together with $A'B'C'$, consider any other triangle $A''B''C''$ inscribed in ABC (Fig. 2.25).

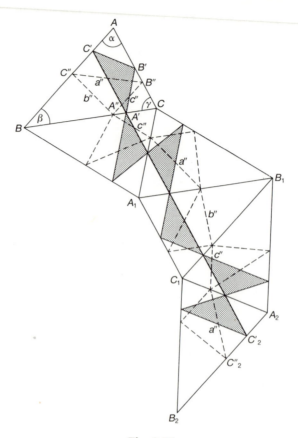

Fig. 2.25

Reflect the whole shape in BC, then reflect the resulting shape in CA_1, then in A_1B_1, then in B_1C_1 and finally in C_1A_2 to obtain triangle $A_2B_2C_1$ (Fig. 2.25).

The side B_2A_2 is parallel to BA, for in the first reflection AB is rotated clockwise through 2β, in the second clockwise through 2α, in the third through $0°$, in the fourth anticlockwise through 2β, and in the fifth reflection anticlockwise through 2α. Thus the total angle through which AB has rotated is $0°$.

Since the sides of $\triangle A'B'C'$ make equal angles with the appropriate sides of $\triangle ABC$, the straight line segment connecting C' with its image C_2' on B_2A_2 is equal to twice the perimeter of $A'B'C'$. The broken line connecting C'' to its image C_2'' on B_2A_2 is twice the perimeter of $\triangle A''B''C''$.

Since the straight line segments $C''C'$ and $C_2''C_2'$ are parallel and congruent, the straight line segment $C'C_2'$ is shorter than the broken line $C''C_2''$. Thus:

The perimeter of $A'B'C'$ is smaller than the perimeter of $A''B''C''$.

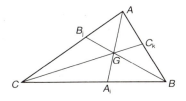

Fig. 2.26

Problem 70

(a) Suppose that for some i, j, k the straight lines AA_i, BB_j and CC_k meet in a point G (Fig. 2.26).

We can consider G as the centre of gravity for a system of masses m_A, m_B, m_C located at A, B and C respectively, such that

> A_i is the centre of gravity of the system consisting of m_B and m_C,
>
> B_j is the centre of gravity of the system consisting of m_A and m_C,
>
> C_k is the centre of gravity of the system consisting of m_A and m_B.

Denote by d_{AB}, d_{BC} and d_{CA} the greatest common divisor of m_A and m_B, m_B and m_C, and m_C and m_A respectively.

Further, put

$$x_{AB} = \frac{m_A + m_B}{d_{AB}}, \qquad x_{BC} = \frac{m_B + m_C}{d_{BC}} \qquad \text{and} \qquad x_{CA} = \frac{m_C + m_A}{d_{CA}}.$$

From mechanics it is known that

$$m_A \cdot AB_j = m_C \cdot B_j C.$$

Since $B_j C = AC - AB_j$, this implies that

$$AB_j = \frac{m_C AC}{m_A + m_C} = \frac{m_C}{d_{CA}} \cdot \frac{AC}{x_{CA}}.$$

Thus, to find B_j we have to divide the side AC into x_{CA} equal parts. Similarly, to find C_k and A_i the sides of AB and BC have to be divided into x_{AB} and x_{BC} equal parts respectively.

According to our problem:

1. All sides of $\triangle ABC$ should be divided into the same number n of equal parts. Hence n must be the least common multiple of x_{AB}, x_{BC} and x_{CA}.
2. n must be odd. Thus x_{AB}, x_{BC} and x_{CA} must be odd. This can happen only if the powers of 2 dividing m_A, m_B and m_C are all different.
3. n should be as small as possible. This will be achieved for $m_A = 2^0$, $m_B = 2^1$ and $m_C = 2^2$.

In that case $x_{AB} = (2^0 + 2^1)/1 = 3$, $x_{BC} = (2^1 + 2^2)/2 = 3$ and $x_{CA} = (2^2 + 2^0)/1 = 5$.

Thus, $n = 15$ is the solution to Problem 70(a).

(b) Problem 70(a) can be extended in the three-dimensional space as follows.

All edges of a tetrahedron $T = ABCD$ are divided into n equal parts. Each division point with the opposite edge of T determines a plane π_i. Find the smallest odd number n such that six of the planes π_i, each through a different edge of T, meet in a point.

The solution can be carried out following the same arguments as in case (a): attach to the vertices of T masses $m_A = 2^0$, $m_B = 2^1$, $m_C = 2^2$ and $m_D = 2^3$. Using the same notation as in (a):

$$x_{AB} = 1 + 2 = 3, \qquad x_{BC} = \frac{2 + 2^2}{2} = 3, \qquad x_{CA} = 1 + 4 = 5,$$

$$x_{AD} = 1 + 2^3 = 9, \qquad x_{BD} = \frac{2 + 2^3}{2} = 5 \quad \text{and} \quad x_{CD} = \frac{2^2 + 2^3}{2^2} = 3.$$

Thus n is the least common multiple of 3, 5 and 9, that is, $n = 45$.

III Problems based on famous topics in the History of Mathematics

Introduction

This chapter presents problems treated by eminent mathematicians in the past. Our selection aims to provide readers with additional information on topics encountered in the school syllabus and to raise interest in the History of Mathematics (see recommended reading).

Chapter III consists of five sections:

1. Problems on prime numbers.
2. The number π.
3. Applications of complex numbers and quaternions.
4. On Euclidean and non-Euclidean geometries.
5. The art of counting.

Part I does not only list the problems; it explains their background and describes solution methods. Beginners are advised to read paragraphs 1.1, 1.3, 2.1, 4.1, 4.2, 5.1 and 5.2, and to attempt solutions of problems marked 'E'.

Part I: Problems

Section 1: Problems on prime numbers

1.1 Prime numbers in arithmetic progressions

The following statement was known already to Euclid, around 300 B.C.:

Theorem 1   There are infinitely many prime numbers.

Theorem 1 can be expressed in the form:

The arithmetic progression 1, 2, 3, . . . contains infinitely many prime numbers.

In the nineteenth century Dirichlet investigated the general question:

Which arithmetic progressions $a, a+d, a+2d, \ldots$, consisting of natural numbers, contain infinitely many primes?

Obviously, if a and d have a common divisor greater than 1, then all terms of the progression, with the possible exception of a, are composite numbers. However, in the remaining case Dirichlet proved:

Theorem 2   Any arithmetic progression $a, a+d, a+2d, \ldots$, in which a and d are relatively prime natural numbers, contains infinitely many primes.

Theorem 2 concerns integers; nevertheless to prove it, Dirichlet used tools of analysis, such as limits and continuity. By doing so, he laid the foundations of analytic number theory — a branch of mathematics in which analysis is applied for the study of integers.

The proof of Theorem 2 is complicated. However, we shall be able to tackle here some of its special cases.

Problem 71 (E)

Prove Theorem 1.

Problem 72

(a) Prove that there are infinitely many prime numbers of the form $4k + 3$, k a positive integer.

  (b) Prove that there are infinitely many prime numbers of the form $4k + 1$, k a positive integer.

1.2 Wilson's theorem and results of Lagrange and Leibniz on prime numbers

In 1770 in his treatise *Meditationes Algebricae*, Waring published a theorem named for his pupil John Wilson:

Theorem 3   For any prime number p the number $(p - 1)! + 1$ is divisible by p.

The first proof of Wilson's theorem was given by Lagrange (in 1773). Lagrange's proof is based on his result on divisibility of values of polynomials by prime numbers.

Theorem 4   Let p be a prime number and let

$$f(x) = a_n x^n + a_{n-1} x^{n-1} + \cdots + a_1 x + a_0$$

be a polynomial of degree $n \geq 1$ with integer coefficients $a_n, a_{n-1}, \ldots, a_0$.

If a_n is not divisible by p, then among the numbers $i = 0, 1, 2, 3, \ldots, p-1$ there are at most n such that $f(i)$ is divisible by p.

Problem 73

Prove Theorem 4.

Theorem 4 has a useful consequence.

Corollary 1   If p is a prime number and if

$$f(x) = a_n x^n + a_{n-1} x^{n-1} + \cdots + a_1 x + a_0$$

is a polynomial with integer coefficients such that for more than n of the integers $i = 0, 1, 2, \ldots, p-1$ the number $f(i)$ is divisible by p, then all coefficients of $f(x)$ are divisible by p.

Problem 74

Prove Collary 1.

Wilson's theorem is deduced from Corollary 1 as follows:

Problem 75

By taking $f(x) = (x-1)(x-2) \cdots (x-p+1) - x^{p-1} + 1$ in Corollary 1, prove Theorem 3.

Wilson's theorem can be applied to establish the following remarkable characterization of prime numbers due to Leibniz (1646–1716):

Theorem 5   A natural number $p > 2$ is a prime if and only if $(p-2)! - 1$ is divisible by p.

Problem 76

Prove Theorem 5 using Wilson's theorem.

1.3 Polynomials with prime number values

Euler (eighteenth century) found a polynomial

$$f(x) = x^2 + x + 41$$

whose values for 40 consecutive integer values of x ($x = 0, 1, 2, \ldots, 39$) are prime numbers. This raises the question:

Problem 77

Find out: Is there a polynomial of degree $m \geq 1$,

$$f(x) = a_m x^m + a_{m-1} x^{m-1} + \cdots + a_1 x + a_0$$

with non-negative integer coefficients a_m, a_{m-1}, . . ., a_0 such that $f(n)$ is a prime number for all natural numbers n?

The next question which comes to mind is:

Are there polynomials with integer coefficients whose set of values, corresponding to non-negative integer variables, contains infinitely many prime numbers?

Obvious examples of such polynomials are $f(x) = x$, or $f(x) = 2x + 1$; Problem 72 points to $f(x) = 4x + 3$ and $f(x) = 4x + 1$. A famous example is the polynomial in two variables $f(x, y) = x^2 + y^2$. In 1650 Fermat formulated the following statement.

Theorem 6   Every prime number of the form $4k + 1$, where k is a positive integer, is the sum of two square numbers.

Theorem 6, combined with Problem 72(b), implies that among the values of $f(x, y) = x^2 + y^2$ there are infinitely many prime numbers.

The proof of Theorem 6 is not difficult, but will not be discussed here. (Interested readers are referred to [60].) Instead we shall consider the following three problems.

Problem 78 (E)

Prove that no prime number of the form $4k + 3$ is the sum of two square numbers.

Problem 79 (E)

Prove that no prime number p of the form $4k + 1$ can be represented as the sum of two squares in two different ways, apart from the order of summands. (That is: If $p = a^2 + b^2$ and $p = c^2 + d^2$, then either $a = c$ and $b = d$, or $a = d$ and $b = c$.)

Problem 80 (E)

Prove that every odd prime number can be represented in a unique way as the difference of two square numbers.

It seems appropriate to conclude the section on prime numbers with a fairly recent result:

In 1976 Matijasevič, Davis, Putnam and Robinson showed how to construct a polynomial $f(x_1, x_2, \ldots, x_n)$ all of whose positive values for non-negative integers $x_1, x_2, \ldots, x_n$ form the set of all prime numbers. The negative values of $f(x_1, x_2, \ldots, x_n)$ are composite numbers.

The above result is impressive. However, if readers happen to think that

$f(x_1, \ldots, x_n)$ represents a neat expression for obtaining primes, they are mistaken!

For more results and open problems on prime numbers see [60], [92].

Section 2: The number π

2.1 Archimedes' algorithm for calculating π

The symbol π for the ratio of the circumference to the diameter in a circle was introduced in the eighteenth century; the ratio itself emerged inevitably in all ancient cultures dealing with geometrical problems. One of the most remarkable early works on estimating π is due to Archimedes of Syracuse (third century B.C.). In his treatise *On the Measurement of the Circle*, Archimedes showed that

$$3\tfrac{10}{71} < \pi < 3\tfrac{10}{70}. \tag{1}$$

The above unusually accurate approximation has not been essentially improved for more than 1800 years. The true significance of Archimedes' result, however, is not in its degree of accuracy but in the ingenious method of calculation.

Archimedes used the fact that the circumference c of a circle lies between the perimeters C_n and I_n of a circumscribed regular n-gon and an inscribed regular n-gon for integral $n \geq 3$. By increasing n we decrease the difference $C_n - I_n$. It follows that by taking sufficiently large values of n we can approximate c with arbitrary accuracy by either C_n or I_n.

The particular achievement of Archimedes was to devise recurrence formulae, expressing C_{2n} and I_{2n} in terms of I_n and C_n:

Problem 81 (E)

Prove that for any $n \geq 3$

(a)
$$C_{2n} = \frac{2C_n I_n}{C_n + I_n}$$

and

(b)
$$I_{2n} = \sqrt{(I_n C_n)}.$$

Having proved this, Archimedes chose the regular hexagon for the start where $C_6 = 4\sqrt{3}r$ and $I_6 = 6r$ (r being the radius of the circle). Using the recurrence formulae, after four steps he found $C_{96} = 3\tfrac{10}{70}$ and $I_{96} = 3\tfrac{10}{71}$, and thus (1).

The method derived by Archimedes is nowadays known as Archimedes' algorithm for calculating π.

2.2 God's delight in odd numbers: the Leibniz series for π, deduced from Gregory's arc tangent series

'Numero deus impari gaudet' — 'God delights in odd numbers!' exclaimed Leibniz in 1674, when he published his discovery on π:

The number $\pi/4$ is the sum of the infinite series of reciprocals of all odd natural numbers with alternating signs:

$$\frac{\pi}{4} = \frac{1}{1} - \frac{1}{3} + \frac{1}{5} - \frac{1}{7} + \cdots . \tag{2}$$

Leibniz's result is a special case of the arc tangent series, found by Gregory in 1671.

Theorem 7   For $0 < x \le 1$

$$\text{arc tan } x = x - \frac{x^3}{3} + \frac{x^5}{5} - \frac{x^7}{7} + \cdots .$$

We shall describe a proof of Theorem 7. This will be done in several steps. (The proof is based on [74].)

Step 1:   Solve the following problem:

Problem 82

If a and b are two numbers such that $a > b \geqslant 0$ prove that

$$\frac{1}{1 + a^2} < \frac{\text{arc tan } a - \text{arc tan } b}{a - b} < \frac{1}{1 + b^2} . \tag{3}$$

(*Hint*: Use Fig. 3.1, where OT is the radius of the unit circle c, t is the tangent to c at T, and A and B are two points on t such that $AT = a$ and $BT = b$.)

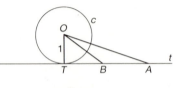

Fig. 3.1

Step 2:   Inequalities (3) reveal a link between arc tan x and the function $f(x)$ $= 1/(1 + x^2)$. Our next task is to determine the mean value of the function $f(x)$ over the interval 0 to x. This is the limit:

$$\lim_{n \to \infty} \frac{f(\delta) + f(2\delta) + \cdots + f(n\delta)}{n} \qquad \text{where } \delta = \frac{x}{n}.$$

Problem 83

Prove that the mean value of μ of $f(x) = 1/(1 + x^2)$ is arc tan x/x.

Step 3: Note that $f(x) = 1/(1 + x^2)$ satisfies the following functional equation:

$$f(x) = 1 - x^2 f(x). \tag{4}$$

Problem 84

(a) By repeated use of (4) show that

$$1 - x^2 + x^4 - \cdots - x^{4n-2} < \frac{1}{1+x^2} < 1 - x^2 + x^4 - x^6 + \cdots + x^{4n}.$$

(b) Deduce that

$$1 - \frac{x^2}{3} + \frac{x^4}{5} - \cdots - \frac{x^{4n-2}}{4n-1} < \frac{\text{arc tan } x}{x} < 1$$
$$- \frac{x^2}{3} + \frac{x^4}{5} - \cdots + \frac{x^{4n}}{4n+1}.$$

Problem 84 leads to

Problem 85

(a) Prove Theorem 7.
(b) Deduce (2) as a special case of Theorem 7.

2.3 π and probability: Buffon's needle problem

In the eighteenth century Buffon started the application of probability to geometrical problems; his work initiated a new branch of probability theory, nowadays called geometrical probability.

It is interesting to note that Buffon was not a mathematician but a biologist, botany being one of his major interests. He was the founder and keeper of the 'Jardin des Plantes', the Paris botanical gardens.

Buffon proposed the following problem:

Suppose that a thin rod is thrown in the air in a room whose floor is made of equal, parallel boards. One of two players bets that the rod will not meet any of the parallel floor joins. The other bets the opposite. Which of the two players has the higher odds?

Buffon suggested playing the game on a checker board with a sewing needle, or a headless pin.

Problem 86

Prove that if the length 2ℓ of the needle is less than or equal to the width $2w$ of the boards, then the probability that the needle cuts a floor join is

$$p = \frac{2\ell}{\pi w}. \tag{5}$$

In the last century formula (5) was used on several occasions for estimating π. In 1850 Wolf in Zurich threw a 36 mm needle 5000 times onto a set of parallel lines 45 mm apart and obtained the value $p \approx 0.5064$ and thus the value $\pi \approx 3.1596$. The Englishman Captain Fox found the value 3.1419 for π, based on 1100 throws.

Section 3: Applications of complex numbers and quaternions

3.1 Gauss' fundamental theorem of axonometry

Imaginary numbers, that is expressions for square roots of negative numbers, made their first appearance as early as the sixteenth century, in the process of solving cubic equations. Nevertheless the foundation of the theory of complex numbers, generalizing the concept of real numbers, was laid only two hundred years later, owing to the work of Gauss.

One of the most celebrated results of Gauss concerning complex numbers, the so-called fundamental theorem of algebra, is about solutions of polynomial equations. It states that:

Any polynomial equation

$$c_n z^n + c_{n-1} z^{n-1} + \cdots + c_1 z + z_0 = 0$$

of degree n with complex coefficients c_i has n (not necessarily distinct) complex (possibly real or imaginary) solutions.

The proof of the above statement will not be given here; interested readers are referred to [82] or [74]. Instead, our aim is to point out that apart from studying complex numbers for their own sake, Gauss introduced them on various occasions for solving problems involving real numbers. Sometimes the use of complex numbers made the solution simpler and sometimes it made formulae more elegant.

We shall describe Gauss' application of complex numbers to axonometry.

Axonometry is a practical discipline. It is concerned with drawing projections of objects on a fixed plane, called the drawing plane π. To facilitate drawing, a three-dimensional coordinate system $\mathbb{C}$ with origin O and coordinate axes Ox, Oy, and Oz is set up in the space. To complete the system we lay off equal line segments OA, OB, and OC on the x, y, and z axes respectively. The structure formed by OA, OB, and OC is called the tripod $OABC$.

Any point P of the object to be projected onto π has coordinates x, y, z with respect to $\mathbb{C}$, and its projection P' is located on π with the help of the tripods projection onto π (Fig. 3.2).

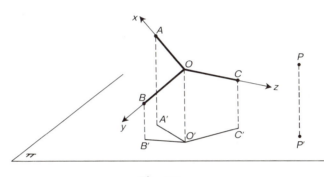

Fig. 3.2

One of the basic problems of axonometry is to answer the following question.

When can three coplanar but non-collinear line segments $O'A'$, $O'B'$, and $O'C'$ represent the orthogonal projection of a tripod?

To solve the above problem Gauss set up a coordinate system in the plane π containing O', A', B' and C', with origin O', and two mutually orthogonal but otherwise arbitrarily chosen straight lines through O' as the x'- and the y'-axes.

By taking the x'-axis as the real axis, and the y'-axis as the imaginary axis, complex numbers were introduced. In this way any point $P'(x', y')$ in the plane was represented by the complex number $c_{P'} = x' + iy'$. In particular, the points A', B', C' were represented by the complex numbers $c_{A'}$, $c_{B'}$, and $c_{C'}$ respectively.

Due to the introduction of complex numbers, the condition for $O'A'B'C'$ to be the orthogonal projection of a tripod $OABC$ was formulated by Gauss in the following convenient, simple form of a quadratic equation:

Theorem 8   If three non-collinear line segments $O'A'$, $O'B'$ and $O'C'$ in a plane π are the orthogonal projections of a tripod, then

$$c_{A'}^2 + c_{B'}^2 + c_{C'}^2 = 0.$$

Theorem 8 is known in the literature as Gauss' fundamental theorem of axonometry.

Problem 87

Prove Theorem 8.

3.2 Lagrange's identity on products of sums of four squares treated by quaternions

It is well known that complex numbers can be used to prove the following statement.

Theorem 9   If two natural numbers n_1, n_2 are expressible as sums of two square numbers, then their product $n_1 n_2$ is also expressible as the sum of two square numbers.

For the proof of Theorem 9, suppose that

$$n_1 = a_1^2 + b_1^2 \quad \text{and} \quad a_2^2 + b_2^2.$$

$\sqrt{(a_1^2 + b_1^2)}$ and $\sqrt{(a_2^2 + b_2^2)}$ can be regarded as the moduli $|z_1|$ and $|z_2|$ of the complex numbers $z_1 = a_1 + ib_1$ and $z_2 = a_2 + ib_2$ respectively.

Since $|z_1| \cdot |z_2| = |z_1 \cdot z_2|$

and

$$z_1 \cdot z_2 = (a_1 + ib_1) \cdot (a_2 + ib_2)$$
$$= (a_1 a_2 - b_1 b_2) + i(a_1 b_2 + a_2 b_1),$$

it follows that

$$\sqrt{(a_1^2 + b_1^2)} \cdot \sqrt{(a_2^2 + b_2^2)} = \sqrt{[(a_1 a_2 - b_1 b_2)^2 + (a_1 b_2 + a_2 b_1)^2]},$$

or

$$(a_1^2 + b_1^2) \cdot (a_2^2 + b_2^2) = (a_1 a_2 - b_1 b_2)^2 + (a_1 b_2 + a_2 b_1)^2.$$

The numbers $a_1 a_2 - b_1 b_2$ and $a_1 b_2 + a_2 b_1$ are real integers. Thus the last equality implies that $n_1 n_2$ is the sum of two square numbers.

Theorem 9 has been extended by Lagrange (eighteenth century) to sums of four square numbers.

Theorem 10 (Lagrange's identity)   If two natural numbers n_1 and n_2 are expressible as sums of four square numbers, then their product $n_1 n_2$ is also expressible as the sum of four square numbers.

Our aim is to describe a proof of Theorem 10, extending the method for the proof of Theorem 9, by introducing quaternions as a generalization of complex numbers.

Quaternions were invented by Hamilton in the nineteenth century as a result of his attempts to describe rotations in three-dimensional space by mathematical formulae.

Hamilton knew that complex numbers $x + iy$, which he regarded as ordered pairs (x, y), can be used to describe rotations in the plane as follows. If P is the point in the complex plane, represented by the complex number $z = x + iy$, then the image P' of P under rotation through θ around the origin O $(0, 0)$ is represented by $z' = e^{i\theta}z$ (where $e^{i\theta} = \cos \theta + i \sin \theta$).

Hamilton's original idea, to treat rotations in space by introducing new numbers represented by ordered triples of real numbers, ended in failure. However, after fifteen years of struggle, Hamilton discovered that rotations in three-dimensional space can be neatly expressed by introducing ordered quadruples of real numbers. Hamilton called these quadruples quaternions. He defined addition and multiplication of quaternions as follows:

Denote the quadruples $(1,0,0,0)$, $(0,1,0,0)$, $(0,0,1,0)$ and $(0,0,0,1)$ by $1, i, j$ and k respectively. Then a quaternion $q = (w, x, y, z)$ can be written as $q = w + xi + yj + zk$.

Addition of two quaternions $q_1 = w_1 + x_1 i + y_1 i + z_1 k$ and $q_2 = w_2 + x_2 i + y_2 i + z_2 k$ is defined 'componentwise', that is

$$q_1 + q_2 = (w_1 + w_2) + (x_1 + x_2)i + (y_1 + y_2)j + (z_1 + z_2)k.$$

The *product* $q_1 q_2$ is calculated using the distributive laws and the associative law:

$$
\begin{aligned}
q_1 q_2 &= (w_1 + x_1 i + y_1 j + z_1 k)(w_2 + x_2 i + y_2 j + z_2 k) \\
&= (w_1 + x_1 i + y_1 j + z_1 k)w_2 + (w_1 + x_1 i + y_1 j + z_1 k)(x_2 i) \\
&\quad + (w_1 + x_1 i + y_1 j + z_1 k)(y_2 j) + (w_1 + x_1 i + y_1 j + z_1 k)(z_2 k) \\
&= w_1 w_2 + x_1 i w_2 + y_1 j w_2 + z_1 k w_2 \\
&\quad + w_1 x_2 i + x_1 i x_2 i + y_1 j x_2 i + z_1 k x_2 i \\
&\quad + w_1 y_2 j + x_1 i y_2 j + y_1 j y_2 j + z_1 k y_2 j \\
&\quad + w_1 z_2 k + x_1 i z_2 k + y_1 j z_2 k + z_1 k z_2 k.
\end{aligned}
\tag{6}
$$

By definition, the products ri, rj, rk are commutative with any real number r, but

$$ij = -ji = k, \qquad jk = -kj = i, \qquad ki = -ik = j;$$

moreover,

$$i^2 = j^2 = k^2 = -1.$$

Thus (6) can be simplified to

$$q_1 q_2 = (w_1 w_2 - x_1 x_2 - y_1 y_2 - z_1 z_2) + (w_1 x_2 + x_1 w_2 + y_1 z_2 - z_1 y_2)i$$
$$+ (w_1 y_2 - x_1 z_2 + y_1 w_2 + z_1 x_2)j + (w_1 z_2 + x_1 y_2 - y_1 x_2 + z_1 w_2)k.$$

The connection between quaternions and rotations is briefly described on p. 225. Here we shall underline the basic significance of quaternions in the history of algebra:

According to Hamilton's definition, the multiplication of quaternions is non-commutative: $ij \neq ji$, $jk \neq kj$, $ki \neq ik$, and, in general, $q_1 q_2$ is not necessarily equal to $q_2 q_1$. The creation of a non-commutative algebraic operation was a revolutionary step in mathematics. It played a vital rôle in the rise of a new branch, the theory of algebraic structures (including groups, fields, rings and vector spaces). For basic properties of algebraic structures see e.g. [61].

Quaternions of the form $w + xi + 0j + 0k$ can be identified with the complex numbers $w + xi$. Thus quaternions represent a generalization of complex numbers, sharing some of their properties. One such property concerns the modulus, or norm of a quaternion, and can be applied to prove Theorem 10.

The norm $|q|$ of a quaternion $q = w + xi + yj + zk$ is the real number $|q| = \sqrt{(w^2 + x^2 + y^2 + z^2)}$.

Problem 88

(a) Prove the following multiplicative property of the norm:

For any quaternions q_1, q_2 the norm of their product is equal to the product of their norms ($|q_1 q_2| = |q_1| |q_2|$).

(b) Deduce the validity of Theorem 10.

Remark: The question arose: Can Theorems 9 and 10 be generalized to sums of n squares? In other words:

For which values of the positive integer n is it true that whenever s_1 and s_2 are sums of n squares, then their product $s_1 s_2$ is also expressible as the sum of n squares — such that the summands of $s_1 s_2$ are bilinear combinations of the summands of s_1 and s_2?

The answer to the above question, given by Hurwitz (nineteenth–twentieth century) is striking: n can be only 1, 2, 4 or 8.

The identity for sums of 8 squares was found by Cayley (nineteenth century).

Section 4: On Euclidean and non-Euclidean geometries

4.1 Euclidean geometry

A mathematical theorem is a mathematical statement whose validity must be established by a rigorous proof. The proof of a theorem has to be deduced from other statements which in their turn have to be proved using earlier statements. Since this process of deduction must start somewhere, any mathematical discipline has to be based on a number of initial assertions — called axioms or postulates — whose truthfulness is accepted without question.

The above was known already to the ancient Greeks. About 300 B.C. Euclid, one of the greatest mathematics teachers of all time, gave an extensive axiomatic treatment of geometry in his celebrated work *Elements*. This was the earliest example of the use of the axiomatic method in the history of mathematics.

In the following centuries Euclid's Postulate 5, nowadays known as *Euclid's parallel postulate*, attracted great attention. In a simplified formulation the postulate states that:

($\mathbb{P}$)   If m is a straight line in a plane π, and P is any point of π not on m, then there is exactly one straight line m' in π passing through P and not intersecting m.

The line m' is called the *parallel* to m through P.

Many reputable geometers believed that ($\mathbb{P}$) is not a postulate but a theorem and tried to prove it. The attempts led to many disputes and controversies. Finally, in the first half of the nineteenth century Bolyai and Lobachevski, independently of each other, constructed a new type of geometry in which Euclid's parallel postulate is *not valid*! In the geometry of Bolyai and Lobachevski ($\mathbb{P}$) is replaced by the following postulate:

($\mathbb{H}$)   In a given plane π, containing an arbitrary point–line pair P, m such that P is not on m, there is more than one line not intersecting m.

The construction of the new geometry, nowadays called *hyperbolic geometry*, can be regarded as one of the greatest mathematical discoveries. It changed the concept of geometry. In the years to follow, a variety of geometries was created, many of them initiating further research and leading to important applications in pure and applied mathematics.

Geometries in which Euclid's parallel postulate ($\mathbb{P}$) is not valid are called non-Euclidean geometries.

In this section we shall describe an ancient problem of Euclidean Geometry, treated by Pappus in *circa* A.D. 320. The problem concerns the 'shoemaker's knife', a configuration studied by Archimedes in the third century B.C.:

Problem 89

a, b, and c_0 are semicircles with diameters AB, AC and CB respectively, such that C is inside the line segment AB, and b and c_0 are inside the semicircle a. The region bounded by a, b, and c_0 is called the *shoemaker's knife* (Fig. 3.3).

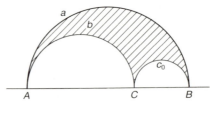

Fig. 3.3

Let c_1, c_2, c_3, . . . be a sequence of circles inscribed in the shoemaker's knife bounded by a, b and c_0 such that c_i touches a, b, and c_{i-1} for $i = 1, 2, 3,$ . . . (Fig. 3.4).

Prove that for all $i \geq 1$ the distance of the centre of c_i from AB is i times the diameter of c_i.

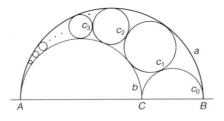

Fig. 3.4

4.2 Projective planes

A *projective plane* is a non-Euclidean geometry, based on just three axioms. These axioms are statements about *points*, *lines*, and *incidence* which are the fundamental notions of this geometry, and hence are not defined:

Axiom 1: Any two distinct points are incident with exactly one common line.

Axiom 2: Any two distinct lines are incident with exactly one common point.

Axiom 3: There exist at least four points such that no three of them are incident with a common line.

If a point *P* is incident with a line ℓ it is customary to say that *P* lies on ℓ, or that ℓ passes through *P*.

There are various types of projective planes. One major difference concerns the number of points (and lines) in a projective plane: this can be finite or infinite. We give an example of a *finite* projective plane (with finitely many points) and of an *infinite* projective plane (with an infinite number of points). It is easy to verify that the geometric structures π_1 and π_2, described below, satisfy Axioms 1–3, and therefore represent projective planes.

Example 1

The points of π_1 are the vertices *A*, *B*, *C*, the feet *D*, *E*, *F* of the altitudes and the centre *G* of the inscribed circle in the equilateral triangle *ABC*. The lines of π_1 are the sides, the altitudes and the inscribed circle. Incidence is 'belonging'; e.g. *D* is incident with the lines *CB*, *AG* and *EF* since it belongs to the side *CB*, to the altitude *AG* and to the inscribed circle (Fig. 3.5).

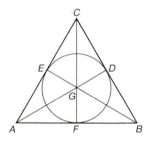

Fig. 3.5

Example 2

π_2 is obtained as the following extension of the classical, Euclidean plane $\mathbb{E}$.

The set of all straight lines in $\mathbb{E}$ is divided into disjoint subsets of mutually parallel lines. Each such subset is called an *ideal point*. The set of all ideal points is called the *ideal line*.

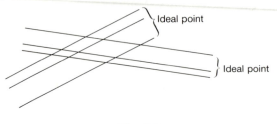

Fig. 3.6

The points of π_2 are the points of $\mathbb{E}$ and the ideal points. The lines of π_2 are the straight lines of $\mathbb{E}$ and the ideal line. Incidence is defined as follows.

A point P of $\mathbb{E}$ is incident with a straight line ℓ of $\mathbb{E}$ if and only if P belongs to ℓ in $\mathbb{E}$; an ideal point I is incident with a straight line ℓ of $\mathbb{E}$ if and only if ℓ belongs to the set of parallel lines of $\mathbb{E}$ representing I; finally, the ideal line is incident with all ideal points and no other point of π_2.

Note that each line in π_1 is incident with the same number of points, each point is incident with the same number of lines, and these two numbers are equal. This property is true for all finite projective planes.

Problem 90

Consider an arbitrary projective plane (finite or infinite). For any point P and any line ℓ denote the set of all lines incident with P by $\{P\}$ and the set of all points incident with ℓ by $\{\ell\}$.
   Prove that:

(a) For any two lines ℓ and ℓ' there is a one-to-one correspondence between the elements of $\{\ell\}$ and $\{\ell'\}$.
       (That is, to each element of $\{\ell\}$ there corresponds a unique element of $\{\ell'\}$ and vice versa.)
(b) For any two points P and P' there is a one-to-one correspondence between the elements of $\{P\}$ and $\{P'\}$.
(c) For any point–line pair P, ℓ there is a one-to-one correspondence between the elements of $\{P\}$ and $\{\ell\}$.

Finite projective planes have the following additional property:

Problem 91

Prove that if in a projective plane the number of points incident with a line is $n + 1$ (where n is a natural number), then the plane consists of $n^2 + n + 1$ points and $n^2 + n + 1$ lines.
   n is called the *order* of the projective plane.

Finite projective planes belong to a special class of combinatorial structures, called balanced incomplete block designs (see Appendix I) and, as such, play an important rôle in finite mathematics and its applications, e.g. in coding theory.

Section 5: The art of counting. Results of Catalan, Euler and André

Counting the number of elements with a given property in a given set is a major task of combinatorics, an important branch of modern mathematics. In this section we shall investigate three famous combinatorial problems on counting from the eighteenth and nineteenth centuries.

5.1 In how many ways can a product of n factors be calculated by pairs?

The above problem was treated by Catalan in 1838. A product is said to be 'calculated by pairs' when always only two factors are multiplied together, and the result of such 'paired multiplication' is used as a factor in the subsequent step of the calculation.

For example, the product of the numbers a, b, c can be calculated by pairs by first forming the product $a \cdot b$ and multiplying the result by c to get $(a \cdot b) \cdot c$. Another way of obtaining the same result is to calculate the product of the pair a, c and then to multiply the result by b; this leads to $(a \cdot c) \cdot b$.

In the products $(a \cdot b) \cdot c$ and $(a \cdot c) \cdot b$ the sequence of the elements a, b, c differs: in $(a \cdot b) \cdot c$ the number a is followed by b and b by c, while in $(a \cdot c) \cdot b$ the number a is followed by c, and c by b. This makes it clear that Catalan's problem consists of two questions:

Question A   How many paired products of n given factors are there if the sequence of the factors is *not* prescribed?

Question B   How many paired products can be constructed from n different factors if the sequence of the factors *is* prescribed?

Denote by R_n the number of paired products of n factors where the sequence of the factors is not prescribed, and by C_n the number of those paired products of n factors where the sequence is prescribed.

Problem 92

Prove that R_n satisfies the recurrence formula

$$R_{n+1} = (4n - 2)R_n.$$

Deduce that

$$R_n = 2 \cdot 6 \cdot 10 \cdot 14 \cdots \cdots (4n - 6) \qquad \text{for } n \geq 2,$$

and

$$C_n = \frac{2 \cdot 6 \cdot 10 \cdot 14 \cdots \cdots (4n - 6)}{n!} \qquad \text{for } n \geq 2.$$

Catalan's problem is interesting in itself.

In 1838 Rodrigues pointed out that Catalan's problem is connected with Euler's problem on polygon division (see 5.2). The solution of the latter caused great difficulties even to Euler. 'The process of induction I employed was quite laborious', he claimed. Rodrigues' idea of using the link between the two topics resulted in a simple solution of Euler's problem.

The discussion of Euler's problem in paragraph 5.2 relies on the following property of C_n.

Problem 93

Prove the following recurrence formula for C_n:

$$C_n = C_1 C_{n-1} + C_2 C_{n-2} + \cdots + C_{n-1} C_1. \tag{7}$$

5.2 Euler's problem on polygon division

In a letter to Goldbach in 1751 Euler posed the following question:

Question C In how many ways can a plane convex polygon of n sides be divided into triangles by diagonals? The dividing diagonals must not cross one another.

Euler knew the answer: The number of different ways of dividing a convex n-gon into triangles by non-crossing diagonals is

$$E_n = \frac{2 \cdot 6 \cdot 10 \cdots \cdots (4n - 10)}{(n - 1)!}.$$

However, Euler was not satisfied with his own method of solution; it involved a laborious process of induction.

In 1758 Segner established a recurrence formula for E_n:

$$E_n = E_2 E_{n-1} + E_3 E_{n-2} + \cdots + E_{n-1} E_2. \tag{8}$$

In 1938 Rodrigues noticed the connection between (7) and (8) and deduced that

$$E_n = C_{n-1}. \tag{9}$$

Problem 94

Prove (8) and (9).

Thus the values of E_n can be easily calculated by combining the results of 5.1 and 5.2.

5.3 The number of 'zigzag' permutations of the set $\{1, 2, 3, \ldots, n\}$ leading to the secant and tangent series

Following André (nineteenth century) let us call a permutation $a_1, a_2, a_3, \ldots, a_n$ of the numbers $1, 2, 3, \ldots, n$ a 'zigzag' permutation if the magnitudes of the permuted numbers alternate, that is either

$$a_1 > a_2 < a_3 > a_4 < a_5 > \cdots$$

or

$$a_1 < a_2 > a_3 < a_4 > a_5 < \cdots$$

André established a recurrence formula for the number Z_n of zigzag permutations of $1, 2, 3, \ldots, n$, and used this result to derive the power series for the functions sec x and tan x.

Our aim in this paragraph is to describe André's work.

A zigzag permutation can begin either by rising ($a_1 < a_2$) or by falling ($a_1 > a_2$) and can end either by rising ($a_{n-1} < a_n$) or by falling ($a_{n-1} > a_n$). It is easy to prove the following.

Problem 95

Denote by A_n, B_n, C_n and D_n the number of those zigzag permutations of $1, 2, \ldots, n$ which begin by rising, begin by falling, end by rising and end by falling respectively.

Prove that

$$A_n = B_n = C_n = D_n = \frac{Z_n}{2}.$$

The next step is to prove the following recurrence formula for A_n.

Problem 96

Prove that

$$2A_n = \sum_{i=0}^{n-1} \binom{n-1}{i} A_i A_{n-1-i}. \tag{10}$$

There is no explicit expression for A_n (or Z_n); however $A_0, A_1, A_2, \ldots$ can be calculated step by step, using (10). So:

$$A_0 = 1, \quad A_1 = 1, \quad A_2 = 1, \quad A_3 = 2, \quad A_4 = 5, \quad A_5 = 16, \ldots .$$

Formula (10) can be simplified by introducing the notation $p_k = A_k/k!$ for $k = 0, 1, 2, \ldots .$

Problem 97

Prove that

$$2np_n = p_0 p_{n-1} + p_1 p_{n-2} + \cdots + p_{n-2} p_1 + p_{n-1} p_0. \tag{11}$$

André spotted a useful connection between formula (11) and the infinite series with coefficients p_i:

$$y = p_0 + p_1 x + p_2 x^2 + \cdots + p_n x^n + \cdots .$$

Namely, one can show that $p_n < \frac{1}{2}$ for all $n \geq 3$. In this case it is well known (see e.g. [62]) that:

1. y converges absolutely, hence y^2 can be expressed in the form

$$y^2 = \sum_{n=1}^{\infty} b_n x^{n-1},$$

where $b_1 = p_0^2 = 1$ and $b_n = p_0 p_{n-1} + p_1 p_{n-2} + \cdots + p_{n-1} p_0$ for $n \geq 2$.
2. y represents a continuous function over every interval $(-h, h)$[1] where $h < 1$, with differentiable terms. Thus y', that is $\mathrm{d}y/\mathrm{d}x$, is equal to

$$y' = \sum_{i=1}^{\infty} i p_i x^{i-1}.$$

---

[1] The interval $(-h, h)$ is the set of real numbers x such that $-h < x < h$.

In view of 1 and 2, and using (11) André could establish a differential equation for y.

Problem 98
(a) Prove that $p_n < \frac{1}{2}$ for all $n \geqslant 3$.
(b) Verify that y satisfies the differential equation

$$1 + y^2 = 2y'. \tag{12}$$

(12) can be easily solved.

Problem 99

Prove that $y = \tan\left(\dfrac{x}{4} + \dfrac{x}{2}\right)$.

Applying standard trigonometric formulae, the following infinite series for sec x and tan x are obtained:

Problem 100
Prove that

$$\sec x = \frac{Z_0}{2} + \frac{Z_2}{2 \cdot 2!} x^2 + \frac{Z_4}{2 \cdot 4!} x^4 + \frac{Z_6}{2 \cdot 6!} x^6 + \cdots;$$
$$\tan x = \frac{Z_1}{2} x + \frac{Z_3}{2 \cdot 3!} x^3 + \frac{Z_5}{2 \cdot 5!} x^5 + \frac{Z_7}{2 \cdot 7!} x^7 \cdots.$$

Part II: Solutions

Section 1: Problems on prime numbers

Problem 71 (E)
Euclid gave the following proof (by contradiction) of Theorem 1.

Suppose that there are only finitely many prime numbers: $p_1, p_2, \ldots, p_n$. Consider the number $N = p_1 \cdot p_2 \cdots \cdot n_n + 1$. The number N cannot be a prime since $N > p_i$ for all $i = 1, 2, \ldots, n$. But if N is a composite number, then N must be divisible by some prime p_i where $1 \leq i \leq n$. Since p_i divides $p_1 \cdot p_2 \cdots \cdot p_n$ it must also divide $N - p_1 \cdot p_2 \cdots \cdot p_n$, which is 1. However, 1 is not divisible by $p_i > 1$. This contradiction shows that there are infinitely many prime numbers.

An alternative proof of Theorem 1, described below, gives the following information about the location of the prime numbers on the number line:

(S)   For any natural number $n > 2$ there is a prime number between n and $n!$

To prove (S) note that the number $M = n! - 1$ has at least one prime divisor p. Clearly, $p < n!$ At the same time, $p > n$; otherwise p would divide $n!$ and also $n! - M$, that is 1, which is impossible.

Problem 72

(a) First we prove the following:

Lemma 1 Any natural number of the form $4\ell + 3$ has at least one prime divisor of the same form.

Proof of Lemma 1   If $4\ell + 3$ is a prime number, then there is nothing to be proved. Otherwise $4\ell + 3 = p_1 \cdot p_2 \cdots \cdot p_r$, where p_i, $i = 1, 2, \ldots, r$, are odd prime numbers, not necessarily different. Suppose that $p_i = 4k_i + 1$ for all $i = 1, 2, \ldots, r$, where the k_i are positive integers.

The product of two numbers $4a + 1$ and $4b + 1$ is equal to

$$(4a + 1)(4b + 1) = 16ab + 4a + 4b + 1 = \underbrace{4(4ab + a + b)}_{} + 1$$
$$= \qquad 4 \cdot c \qquad \cdot + 1.$$

Hence

$$(4k_1 + 1) \cdot (4k_2 + 1) \cdots \cdots (4k_r + 1) = 4t + 1$$

for some integer t. The number $4t + 1$ cannot be equal to $4\ell + 3$. This contradiction shows that $4\ell + 3$ must have a prime divisor of the form $4k + 3$.

Now we are ready to prove that there are infinitely many prime numbers of the form $4k + 3$.

Consider $N = 4n! - 1$ where n is an arbitrary natural number. N is of the form $4\ell + 3$, hence, according to Lemma 1, it has a prime divisor p of the form $4k + 3$. Suppose that $p \leq n$. In that case p divides $4n!$ Hence p must divide $N - 4n!$, that is p must divide 1, which is impossible. This implies that $p > n$. Thus for any natural number n there exists a prime $p = 4k + 3 > n$. This proves that there are infinitely many prime numbers of the form $4k + 3$.

(b) There are infinitely many prime numbers of the form $4k + 1$.

Following Sierpiński [91] we shall deduce the proof of the above statement from Fermat's little theorem (Problem 45) combined with the well-known formula:

$$a^{2t+1} + 1 = (a + 1)(a^{2t} - a^{2t-1} + a^{2t-2} - \cdots - a + 1). \qquad (13)$$

We start with the following lemma.

Lemma 2   For any natural number $n > 1$ all prime divisors of the number $M = (n!)^2 + 1$ are of the form $4k + 1$.

Proof of Lemma 2. M is an odd number, therefore all its prime divisors are odd. Suppose that M has a prime divisor of the form $p = 4k + 3$. Obviously, $p > n$. According to Fermat's little theorem:

$$p \text{ divides } (n!)^p - n! \tag{14}$$

Our aim is to show that:

$$\text{If } p = 4k + 3, \qquad \text{then } p \text{ divides } (n!)^p + n! \tag{15}$$

To prove (15), apply (13) to the special case when $a = (n!)^2$ and $t = k$. In that case $a + 1 = M$. Hence, according to (13), M divides $[(n!)^2]^{2k+1} + 1$. But

$$[(n!)^2]^{2k+1} + 1 = (n!)^{4k+2} + 1 = (n!)^{p-1} + 1.$$

Thus M divides $(n!)^{p-1} + 1$ and therefore also divides

$$n![(n!)^{p-1} + 1] = (n!)^p + n!$$

Combining (14) and (15) one deduces that p divides $2(n!)^p$, that is p divides $2n!$

No prime divisor of $2n!$ is greater than n. However, $p > n$; this contradiction proves Lemma 2.

According to Lemma 2, for any natural number $n > 1$ there is a prime number $p = 4k + 1$ greater than n. Thus there are infinitely many prime numbers of the form $4k + 1$.

Problem 73

Theorem 4 will be proved by induction on n.

Step 1:   Let $n = 1$ and suppose that there are two integers x_1, x_2 such that $0 \le x_1 < x_2 \le p - 1$ and that p divides $f(x_1)$ and $f(x_2)$. Since $f(x) = a_1 x + a_0$, this implies that p divides $f(x_2) - f(x_1) = a_1(x_2 - x_1)$. Thus p must divide at least one of the factors a_1 and $x_2 - x_1$. However, $x_2 - x_1 < p$ and a_1 is relatively prime to p, according to the assumption of Theorem 4.

This contradiction shows that Theorem 4 is true for $n = 1$.

Step 2:   Suppose that Theorem 4 is true for $n - 1$ but not for n. Thus there are $n + 1$ integers $x_1, x_2, \ldots, x_{n+1}$ such that

$$0 \le x_1 < x_2 < \cdots < x_{n+1} < p$$

and p divides $f(x_i)$ when $f(x)$ is of degree n. Form the difference:

$$f(x) - f(x_1) = a_n(x^n - x_1^n) + a_{n-1}(x^{n-1} - x_1^{n-1}) + \cdots + a_1(x - x_1). \quad (16)$$

Since

$$x^k - x_1^k = (x - x_1)(x^{k-1} + x^{k-2}x_1 + \cdots + xx_1^{k-2} + x_1^{k-1})$$

equation (16) can be written as

$$f(x) - f(x_1) = (x - x_1)g(x),$$

where $g(x)$ is a polynomial of degree $n - 1$ with leading coefficient a_n.

$f(x_i) - f(x_1)$ is divisible by p for $i = 2, 3, \ldots, n+1$. Hence p divides $(x_i - x_1)g(x_i)$ for $i = 2, 3, \ldots, n+1$. The number $x_i - x_1$ is less than p; therefore p must divide $g(x_i)$ for $i = 2, 3, \ldots, n+1$, contradicting the induction hypothesis.

This finishes the proof of Theorem 4.

Problem 74

Corollary 1 can be proved by contradiction.

Suppose that $f(x)$ satisfies the conditions of the statement, and that not all coefficients a_i are divisible by p. Let a_k be the first coefficient in the sequence $a_n, a_{n-1}, \ldots, a_{k+1}, a_k, \ldots, a_1, a_0$ which is not divisible by p.

Form the polynomial

$$g(x) = f(x) - (a_n x^n + a_{n-1}x^{n-1} + \cdots + a_{k+1}x^{k+1})$$
$$= a_k x^k + \cdots + a_0.$$

For all $x_i \in \{0, 1, \ldots, p-1\}$ such that p divides $f(x_i)$ the numbers $g(x_i)$ are also divisible by p. There are more than n such numbers $g(x_i)$. The polynomial $g(x)$ is of degree $k \le n$; thus, according to Theorem 4, its degree k must be 0. This implies that

$$a_n, a_{n-1}, \ldots, a_1 \text{ are all divisible by } p. \quad (17)$$

On the other hand,

$$a_0 = f(x) - (a_n x^n + \cdots + a_1 x). \quad (18)$$

Choose $x_i \in \{0, 1, \ldots, p-1\}$ such that p divides $f(x_i)$. From (17) and (18) it follows that a_0 is divisible by p.

This proves Corollary 1.

Problem 75 (Proof of Wilson's theorem)

Consider the polynomial

$$f(x) = (x - 1)(x - 2) \cdot \cdots \cdot (x - p + 1) - x^{p-1} + 1.$$

$f(x)$ is of degree $p - 2$ and îts neutral term a_0 is equal to $(p - 1)! + 1$. Our aim is to show that there are at least $p - 1$ values $x_i \in \{0, 1, . . ., p-1\}$ such that p divides $f(x_i)$. This will enable us to show, using Corollary 1, that p divides $(p - 1)! + 1$.

Put $x_i = i$ for $i = 1, 2, . . ., p-1$. The expression $(x_i - 1)(x_i - 2) \cdots$ $(x_i - p + 1)$ is 0 for all $i = 1, 2, . . ., p - 1$. Since 0 is divisible by p this implies that

$$(x_i - 1)(x_i - 2) \cdots (x_i - p + 1) \text{ is divisible by } p. \qquad (19)$$

According to Fermat's little theorem $x_i^p - x_i = x_i(x_i^{p-1} - 1)$ is divisible by p. The numbers x_i and p are coprime, hence

$$x_i^{p-1} - 1 \text{ is divisible by } p. \qquad (20)$$

Combining (19) and (20) we see that

$$f(x_i) \text{ is divisible by } p \qquad \text{for } i = 1, 2, . . ., p-1.$$

Hence, by Corollary 1, all coefficients of $f(x)$ are divisible by p. In particular:

$$(p - 1)! + 1 \text{ is divisible by } p.$$

Problem 76 (Proof of Leibniz' theorem)

(a) Suppose that $p > 2$ is a prime number. Then, according to Wilson's theorem, p divides $(p - 1)! + 1$. But

$$(p - 1)! + 1 = (p - 2)!(p - 1) + 1$$
$$= (p - 2)!p - [(p - 2)! - 1].$$

Hence p divides $(p - 2)! - 1$.

(b) Let $p > 2$ be a natural number dividing $(p - 2)! - 1$. Then p also divides $(p - 1)[(p - 2)! - 1]$ which is equal to $(p - 1)! - p + 1$. Thus

$$p \text{ divides } (p - 1)! + 1.$$

Suppose that p is not a prime. Then $p = a \cdot b$ for some integers a, b greater than 1 and less than p. In that case a divides both $(p-1)! + 1$ and $(p-1)!$, which is impossible since a does not divide 1.

Hence p must be a prime.

Problem 77

Choose a natural number n_0 such that $f(n_0) = a > 1$ and consider $f(n_0 + ka)$ for any natural number k. In the polynomial

$$g(n_0 + ka) = f(n_0 + ka) - f(n_0)$$

$$= a_m[(n_0 + ka)^m - n_0^m] + a_{m-1}[(n_0 + ka)^{m-1} - n_0^{m-1}] + \cdots$$

$$+ \cdots + a_1[(n_0 + ka) - n_0]$$

every difference $(n_0 + ka)^i - n_0^i$ is divisible by $(n_0 + ka) - n_0 = ka$, hence also by a.

Thus $g(n_0 + ka)$ is divisible by a. Since $f(n_0) = a$ this implies that $f(n_0 + ka)$ is divisible by a. From $f(n_0 + ka) > a$ and $a > 1$ it follows that $f(n_0 + ka)$ is not a prime by any $k = 1, 2, 3, \ldots$.

Problem 78 (E)

Suppose that $n = 4k + 3 = a^2 + b^2$ for two natural numbers a and b. Since n is odd, one of the numbers, say a, must be even, and the other number b must be odd. Thus $a = 2a_1$ and $b = 2b_1 + 1$ for some a_1 and b_1. Hence

$$a^2 + b^2 = (2a_1)^2 + (2b_1 + 1)^2 = 4(a_1^2 + b_1^2 + b_1) + 1.$$

This implies that n is of the form $4k + 1$ and not of the form $4k + 3$ — a contradiction.

Problem 79 (E)

Suppose that $p = a^2 + b^2 = c^2 + d^2$. In that case

$$p^2 = (a^2 + b^2)(c^2 + d^2).$$

Hence p^2 can be expressed in the following two ways:

$$p^2 = (ac + bd)^2 + (ad - bc)^2 \tag{21}$$

or

$$p^2 = (ac - bd)^2 + (ad + bc)^2. \tag{22}$$

Moreover,

$$(ac+bd)(ad+bc) = (a^2+b^2)cd + (c^2+d^2)ab = p(cd+ab). \qquad (23)$$

From (23) it follows that p divides at least one of the numbers $ac + bd$, $ad + bc$.

Case 1: Suppose that p divides $ac + bd$. Then, in view of (21), $p = ac + bd$ and $ad - bc = 0$, that is $ad = bc$. Hence

$$pa = a^2c + bad = a^2c + b^2c = (a^2 + b^2)c = pc.$$

The last equation implies that $a = c$. Thus $b = d$.

Case 2: If p divides $ad + bc$, then (22) implies that $p = ad + bc$ and $ac = bd$. In that case

$$pa = a^2d + bac = a^2d + b^2d = (a^2 + b^2)d = pd.$$

Thus $a = d$ and $b = c$.

Problem 80 (E)

Our aim is to find two natural numbers a and b such that $p = a^2 - b^2$, that is

$$p = (a + b)(a - b).$$

Since p is a prime, the above equation implies that

$$a + b = p \quad \text{and} \quad a - b = 1.$$

Thus

$$a = \frac{p + 1}{2} \quad \text{and} \quad b = \frac{p - 1}{2}.$$

p is an odd number greater than 1, hence $(p + 1)/2$ and $(p - 1)/2$ are natural numbers. In other words, p can be expressed as the difference of two square numbers in exactly one way:

$$p = \left(\frac{p + 1}{2}\right)^2 - \left(\frac{p - 1}{2}\right)^2.$$

Section 2: The number π

Problem 81

In Fig. 3.7 the point O is the centre of a circle c; AB is the side of an inscribed regular n-gon in c, and AC and CB are sides of an inscribed regular $2n$-gon. EF is the side of a circumscribed $2n$-gon and AG is half the side of a circumscribed regular n-gon. The chord DC is equal to AB.

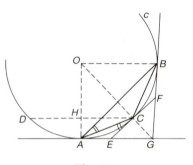

Fig. 3.7

Put $AB = i_n$, $AC = i_{2n}$, $EF = c_{2n}$ and $AG = \frac{1}{2}c_n$.

(a) To express c_{2n} in terms of c_n and i_n, consider the following two pairs of similar triangles: EGC and OGA, and OHC and OAG.

Comparison of their sides leads to the proportions:

$$EC : EG = OA : OG = OC : OG = HC : AG,$$

that is

$$\frac{c_{2n}}{2} : \left(\frac{c_n}{2} - \frac{c_{2n}}{2}\right) = \frac{i_n}{2} : \frac{c_n}{2}.$$

This implies that

$$c_{2n} = \frac{c_n i_n}{c_n + i_n},$$

or

$$2nc_{2n} = \frac{2n\, c_n \cdot n i_n}{n c_n + n i_n}. \tag{24}$$

$2nc_{2n}$, nc_n, and ni_n are the perimeters C_{2n}, C_n, I_n of the circumscribed $2n$-gon, circumscribed n-gon, and inscribed n-gon respectively. Thus (24) can be rewritten as

$$C_{2n} = \frac{2C_n I_n}{C_n + I_n}.$$

(b) i_{2n} can be expressed in terms of i_n and c_{2n} by considering the similar triangles AEC and ACB: The proportion $AE : AC = AC : AB$ leads to

$$i_{2n} = \sqrt{\left(\frac{c_{2n}}{2} \cdot i_n\right)},$$

that is, to

$$2ni_{2n} = \sqrt{(2nc_{2n} \cdot ni_n)}.$$

In other words

$$I_{2n} = \sqrt{(C_{2n}I_n)}.$$

Problem 82

Consider Fig. 3.8. The area of triangle OAB is between the areas of the circle sectors OBB' and $OA'A$, that is

$$\tfrac{1}{2}OB^2(\alpha - \beta) < \tfrac{1}{2}BA \cdot 1 < \tfrac{1}{2}OA^2(\alpha - \beta). \tag{25}$$

Since

$$OB^2 = 1 + b^2,\ OA^2 = 1 + a^2,\ BA = a - b,$$

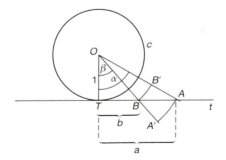

Fig. 3.8

and

$$\alpha - \beta = \arctan \frac{a}{1} - \arctan \frac{b}{1}$$

the inequalities (25) imply the inequalities

$$\frac{1}{1 + a^2} < \frac{\arctan a - \arctan b}{a - b} < \frac{1}{1 + b^2}. \tag{26}$$

Problem 83

Put $f(x) = 1/(1 + x^2)$. Divide the interval $[0, x]^1$ into n equal parts $\delta = x/n$. Write down inequalities (26) for $a = i\delta$, $b = (i - 1)\delta$, $i = 1, 2, \ldots, n$, and add them together. This gives

$$\sum_{i=1}^{n} f(i\delta) < \frac{\arctan x}{\delta} < \sum_{i=1}^{n} f((i - 1)\delta) = \sum_{i=0}^{n-1} f(i\delta). \tag{27}$$

After division by n, the inequalities (27) yield the inequalities

$$\frac{1}{n} \sum_{i=1}^{n} f(i\delta) < \frac{\arctan x}{x} < \frac{1}{n} \sum_{i=1}^{n} f(i\delta) + \frac{1}{n}\left(1 - \frac{1}{1 + x^2}\right). \tag{28}$$

Since

$$\lim_{n \to \infty} \frac{1}{n}\left(1 - \frac{1}{1 + x^2}\right) = 0,$$

the inequalities (28) imply that the mean value μ of $f(x) = 1/(1 + x^2)$ is

$$\mu = \frac{\arctan x}{x}.$$

Problem 84

(a) By induction it can be proved that

$$\frac{1}{1 + x^2} = 1 - x^2 + x^4 - x^6 + \cdots + (-1)^k x^{2k} \frac{1}{1 + x^2}$$

for $k = 0, 1, 2, \ldots$.

---

^{1} $[0, x]$ is the set of all real numbers r such that $0 \le r \le x$.

Hence

$$1 - x^2 + \cdots + (-1)^{2n-1}x^{4n-2} < \frac{1}{1 + x^2}$$
$$< 1 - x^2 + x^4 - \cdots + (-1)^{2n}x^{4n}. \tag{29}$$

(b) Denote the sum $1 - x^2 + \cdots + (-1)^k x^{2k}$ by $S_k(x)$, and its mean value over $[0, x]$ by μ_k. In view of (29) the mean value μ of $1/(1 + x^2)$ over $[0, x]$ lies between μ_{2n-1} and μ_{2n}:

$$\mu_{2n-1} < \mu < \mu_{2n}. \tag{30}$$

According to the definition of the mean value, if $\delta = x/n$,

$$\mu_k = \lim_{n \to \infty} \frac{1}{n} \sum_{i=0}^{k} (-1)^i [\delta^{2i} + (2\delta)^{2i} + \cdots + (n\delta)^{2i}].$$

The sum $\delta^{2i} + (2\delta)^{2i} + \cdots + (n\delta)^{2i}$ can be rewritten as

$$\delta^{2i}(1^{2i} + 2^{2i} + \cdots + n^{2i}) = \delta^{2i}S_{2i, n},$$

where $S_{2i, n} = 1^{2i} + 2^{2i} + \cdots + n^{2i}$ is a polynomial in n of degree $2i + 1$ with leading coefficient $1/(2i + 1)$ (see Problem 66). Thus

$$\lim_{n \to \infty} \frac{(-1)^i(\delta^{2i} + (2\delta)^{2i} + \cdots + (n\delta)^{2i}}{n} = \lim_{n \to \infty} \frac{(-1)^i\delta^{2i}}{n} \cdot \frac{n^{2i+1}}{2i + 1}$$

$$= \lim_{n \to \infty} \frac{(-1)^i(x/n)^{2i}}{n} \cdot \frac{n^{2i+1}}{2i + 1}$$

$$= \frac{(-1)^i x^{2i}}{2i + 1}.$$

Hence

$$\mu_k = 1 - \frac{x^2}{3} + \frac{x^4}{5} - \cdots + (-1)^k \frac{x^{2k}}{2k + 1}. \tag{31}$$

The mean value of $1/(1 + x^2)$ is (arc tan $x)/x$ (see Problem 83); thus (30) and (31) imply that

$$1 - \frac{x^2}{3} + \frac{x^4}{5} - \cdots - \frac{x^{4n-2}}{4n-1} < \frac{\arctan x}{x} < 1$$

$$- \frac{x^2}{3} + \frac{x^4}{5} - \cdots + \frac{x^{4n}}{4n+1}. \tag{32}$$

Problem 85

(a) From (32) it follows that

$$x - \frac{x^3}{3} + \frac{x^5}{5} - \cdots - \frac{x^{4n-1}}{4n-1} < \arctan x < x$$

$$- \frac{x^3}{3} + \frac{x^5}{5} - \cdots + \frac{x^{4n+1}}{4n+1}. \tag{33}$$

The difference

$$d_n = x - \frac{x^3}{3} + \frac{x^5}{5} - \cdots + \frac{x^{4n+1}}{4n+1} - \arctan x$$

is smaller than the difference

$$d'_n = x - \frac{x^3}{3} + \frac{x^5}{5} - \cdots + \frac{x^{4n+1}}{4n+1}$$

$$- \left(x - \frac{x^3}{3} + \frac{x^5}{5} - \cdots - \frac{x^{4n-1}}{4n-1} \right)$$

$$= \frac{x^{4n+1}}{4n+1}.$$

If $0 < x \le 1$, then $\lim_{n \to \infty} d'_n = 0$, therefore also $\lim_{n \to \infty} d_n = 0$. Hence (33) leads to Gregory's arc tangent series:

$$\arctan x = x - \frac{x^3}{3} + \frac{x^5}{5} - \frac{x^7}{7} + \cdots$$

(b) Substitution of $x = 1$ in Gregory's series yields Leibniz' formula for π:

$$\frac{\pi}{4} = 1 - \frac{1}{3} + \frac{1}{5} - \frac{1}{7} + \cdots.$$

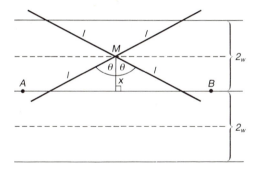

Fig. 3.9

Problem 86

For an arbitrary position of the needle on the floor denote by x the distance of its midpoint M from a board joint AB, and by θ the angle between the needle and the perpendicular to AB (Fig. 3.9). The needle will intersect AB if and only if $-\ell \leq x \leq \ell$ and $-\theta(x) \leq \theta \leq \theta(x)$, where $\theta(x) = \text{arc cos } (x/\ell)$.

The needle is thrown at random; hence x and θ are uniformly distributed. Thus

the probability that M falls in the interval $(x, x+\Delta x)$ is $\Delta x/(2w)$ (34)

and

the probability that the angle between the needle and the perpendicular is between θ and $\theta + \Delta$ is $\Delta\theta/\pi$. (35)

(35) implies that the probability of the needle intersecting AB for given x is

$$p_x = \int_{-\theta(x)}^{\theta(x)} \frac{d\theta}{\pi} = \frac{2\theta(x)}{\pi}.$$

Hence, the probability that the needle intersects AB is

$$p = \int_{x=-\ell}^{x=\ell} p_x \cdot \text{probability of } M \text{ falling in } (x, x+\Delta x)$$

$$= \int_{-\ell}^{\ell} \frac{2\theta(x)}{\pi} \cdot \frac{\mathrm{d}x}{2w} = 2 \int_0^\ell \frac{\theta(x)\,\mathrm{d}x}{\pi w} = 2 \int_0^\ell \frac{\mathrm{arc\ cos}\ (x/\ell)}{\pi w}\,\mathrm{d}x$$

$$= \frac{2\ell}{\pi w}. \tag{36}$$

From (36) it follows that the probability of the needle intersecting AB is greater than $\frac{1}{2}$ if and only if $2\ell/(\pi w) > \frac{1}{2}$, that is $\ell/w > \pi/4 > 0.78539$. This means that if the length of the needle were, say, $\frac{4}{5}$ of the floorboard width the odds would be in favour of intersecting a join. On the other hand, if the needle's length were, say, $\frac{3}{4}$ of the width of the floorboard the odds would be against intersecting the join.

Section 3: Applications of complex numbers and quaternions

Problem 87 (Gauss' fundamental theorem of axonometry)

Suppose that $O'A'$, $O'B'$ and $O'C'$ are the normal projections of the legs OA, OB and OC of a tripod $OABC$ onto a plane π. Take the common length of OA, OB and OC as our unit length.

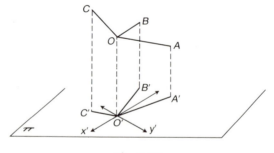

Fig. 3.10

Denote the angles of OA, OB and OC with the x'-axis by α, β and γ respectively, and the angles of OA, OB and OC with the y'-axis by α', β' and γ' respectively. Thus:

$$c_{A'} = \cos \alpha + \mathrm{i} \cos \alpha', \; c_{B'} = \cos \beta + \mathrm{i} \cos \beta',$$

$$c_{C'} = \cos \gamma + \mathrm{i} \cos \gamma',$$

and

$$c_{A'}^2 + c_{B'}^2 + c_{C'}^2 = (\cos \alpha + i \cos \alpha')^2 + (\cos \beta + i \cos \beta')^2$$
$$+ (\cos \gamma + i \cos \gamma')^2$$
$$= (\cos^2\alpha + \cos^2\beta + \cos^2\gamma)$$
$$+ 2i(\cos \alpha \cos \alpha' + \cos \beta \cos \beta' + \cos \gamma \cos \gamma')$$
$$- (\cos^2\alpha' + \cos^2\beta' + \cos^2\gamma').$$

$\cos \alpha$, $\cos \beta$ and $\cos \gamma$ are the direction cosines of the x'-axis in the coordinate system $\mathbb{C}$ formed by OA, OB and OC. Since OA, OB and OC are mutually perpendicular,

$$\cos^2\alpha + \cos^2\beta + \cos^2\gamma = 1.$$

Similarly, $\cos \alpha'$, $\cos \beta'$ and $\cos \gamma'$ are the direction cosines of the y'-axis in $\mathbb{C}$, thus

$$\cos^2\alpha' + \cos^2\beta' + \cos^2\gamma' = 1.$$

Since $\cos \alpha \cos \alpha' + \cos \beta \cos \beta' + \cos \gamma \cos \gamma'$ is the scalar product of the unit vectors on the mutually orthogonal x'- and y'-axes, it equals 0.

The above arguments imply that $c_{A'}^2 + c_{B'}^2 + c_{C'}^2 = 1 + 0 - 1 = 0$. This proves Gauss' theorem.

Problem 88

(a) If $q_1 = w_1 + x_1 i + y_1 j + z_1 k$ and $q_2 = w_2 + x_2 i + y_2 j + z_2 k$, then

$$q_1 q_2 = w_3 + x_3 i + y_3 j + z_3 k,$$

where

$$\left. \begin{array}{l} w_3 = w_1 w_2 - x_1 x_2 - y_1 y_2 - z_1 z_2 \\ x_3 = w_1 x_2 + x_1 w_2 + y_1 z_2 - z_1 y_2 \\ y_3 = w_1 y_2 - x_1 z_2 + y_1 w_2 + z_1 x_2 \\ z_3 = w_1 z_2 + x_1 y_2 - y_1 x_2 + z_1 w_2 \end{array} \right\} \tag{37}$$

It is a matter of straightforward calculations to check that

$$\sqrt{(w_3^2 + x_3^2 + y_3^2 + z_3^2)} = \sqrt{(w_1^2 + x_1^2 + y_1^2 + z_1^2)} \cdot \sqrt{(w_2^2 + x_2^2 + y_2^2 + z_2^2)}.$$

(b) Suppose that for $t = 1, 2,$

$$n = w_t^2 + x_t^2 + y_t^2 + z_t^2, \text{ where } w_t, x_t, y_t, \text{ and } z_t \text{ are integers.} \qquad (38)$$

Consider the quaternions $q_t = w_t + x_t i + y_t j + z_t k$ with norm $\sqrt{n_t}$ for $t = 1, 2$. According to (a):

$$n_1 n_2 = |q_1 q_2|^2 = w_3^2 + x_3^2 + y_3^2 + z_3^2.$$

From (37) and (38) it follows that w_3, x_3, y_3 and z_3 are integers. Thus $n_1 n_2$ is the sum of four squares.

Section 4: On Euclidean and non-Euclidean geometries

Problem 89

This problem can be solved using inversion with respect to the circle c with centre A, orthogonal to c_i for a fixed $i \in \{1, 2, 3, \ldots\}$.

Recall that (see Appendix I)

Inversion with respect to a circle c with centre O and radius r is a transformation mapping an arbitrary point $P \neq 0$ in the plane of c onto the point P' such that:

P' is on the straight line OP, on the same side of O as P,

and

the product of the distances OP and OP' is
$$OP \cdot OP' = r^2.$$

(Fig. 3.11).

It is left to the reader to verify that in the plane of c:
- every point P except O has a unique image P';
- $(P')' = P$;
- the points inside c are mapped onto points outside c and vice versa (hence the name 'inversion': the inside of c is 'inverted' to its outside);

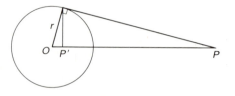

Fig. 3.11

- the points of c remain fixed;
- any straight line through O is mapped onto itself;
- any straight line not through O is mapped onto a circle through O, and vice versa;
- any circle not through O is mapped onto a circle not through O, and
- any circle orthogonal to c is mapped onto itself.

    (Two circles c and c' are called orthogonal if they intersect at right angles, that is, if their tangents at the points of intersection make a right angle.)

Now consider the shoemaker's knife bounded by the semi-circles a, b and c_0, and the sequence of circles $c_1, c_2, c_3, \ldots$ inscribed in it so that c_i touches a, b and c_{i-1} for $i = 1, 2, \ldots$ (Fig. 3.12). Construct the image of this configuration under the inversion α with respect to the circle c with centre A, orthogonal to c_i for some $i \geq 1$.

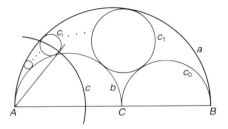

Fig. 3.12

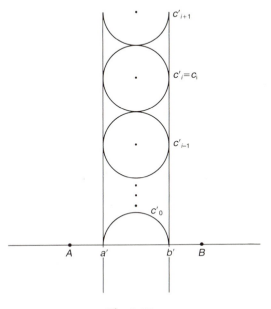

Fig. 3.13

α leaves c_i and the line AB fixed. It maps the circles a and b onto the straight lines a' and b' respectively, tangent to c_i and orthogonal to AB (Fig. 3.13).

The circles $c_0, c_1, \ldots, c_{i-1}$ are mapped onto circles $c_0', c_1', \ldots, c_{i-1}'$, all touching a' and b'. Moreover, c_k' touches c_{k-1}' for $k = 1, \ldots, i-1$ and c_{i-1}' touches $c_i' = c_i$. The centre of c_0 is on AB. The diameters of the circles $c_0', c_1', c_{i-1}', \ldots, c_i'$ are all equal to the distance d_i between a' and b'. Thus the distance of the centre O_i' of c_i' from AB is equal to id_i. Since $c_i' = c_i$ this implies that d_i and O_i' are respectively the diameter and the centre of c_i. Thus the distance of the centre of c_i from AB is i times the diameter of c_i. This is true for any $i = 1, 2, 3, \ldots$, since i was chosen arbitrarily.

Problem 90

Our first step is to show that:

1. for any two lines ℓ, ℓ' of a projective plane π there is a point M in π not incident with either ℓ or ℓ';
2. for any two points P, P' of π there is a line g not incident with either P or P'.

To show this recall that, according to Axiom 3, there are four points, say A, B, C, D in π, no three of them incident with a common line. These points determine three pairs of lines: AB and CD, BC and AD, and AC and BD. Denote the point incident with AB and CD by R, the point incident with BC and AD by S, and the point incident with AC and BD by T (Fig. 3.14).

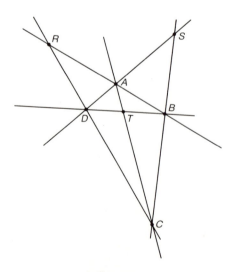

Fig. 3.14

It is easy to verify that
- for any two lines ℓ, ℓ' at least one of the points A, B, C, D, R, S, T is not incident with either of them, and that
- for any two points P, P' at least one of the lines AB, CD, BC, AD, AC, BD is not incident with either.

This proves (1) and (2).

It is now easy to prove (a), (b) and (c):

(a) Let ℓ and ℓ' be two arbitrary distinct lines of π. Let A be a point of π not incident with either ℓ or ℓ'. Using Axioms 1 and 2 we associate with an arbitrary point M of $\{\ell\}$ a point M' of $\{\ell'\}$ as follows: M' is the point incident with both ℓ' and AM, where AM is the line incident with A and M. The mapping $M \to M'$ establishes a one-to-one correspondence between $\{\ell\}$ and $\{\ell'\}$ (Fig. 3.15).

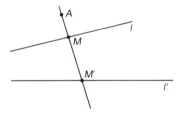

Fig. 3.15

(b) In the text of the proof of (a) replace the lines ℓ, ℓ' and the points A, M, and M' by the points P, P' and by the lines a, m, and m' respectively. Also interchange the words 'point' and 'line'. The resulting text is a proof of (b).

Remark: The statement:
> If in any valid proposition S about incidence of points and lines in a projective plane the words 'point' and 'line' are interchanged, while the remaining text is left unaltered, then a valid proposition S' is obtained.

is called the *principle of duality*, in π, and S' is called the dual of S.

The validity of the principle of duality in π is a consequence of the fact that Axiom 2 is the dual of Axiom 1 and the dual of Axiom 3 is a valid proposition in π.

(c) If P is not incident with ℓ, then with any point $M \in \{\ell\}$ associate the line PM incident with both P and M. The mapping $M \to PM$ establishes a one-to-one correspondence between $\{\ell\}$ and $\{P\}$ (Fig. 13.16(a)). If P is incident with ℓ, choose a line ℓ' not incident with P, and a point A not incident with either ℓ or ℓ'. With any point $X \in \{\ell\}$ associate the line AX (incident with A and X), with AX associate the point $X' \in \{\ell'\}$ incident

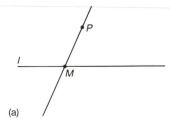

Fig. 3.16(a)

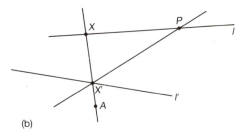

Fig. 3.16(b)

with AX, and with X' associate the line PX' of $\{P\}$ (incident with P and X') (Fig. 3.16(b)). The sequence of mappings $X \rightarrow AX \rightarrow X' \rightarrow PX'$ establishes a one-to-one correspondence between $\{\ell\}$ and $\{P\}$.

Problem 91

According to the statements proved in Problem 90, if one line of π is incident with n + 1 points, then all lines of π are incident with n + 1 points, and all points of π are incident with n + 1 lines.

Let P be an arbitrary point of π. Any point of π different from P is incident with exactly one of the n + 1 lines through P (Axiom 1). Together, the n + 1 lines through P carry (n + 1)n points different from P. Thus the total number of points in π is (n + 1)n + 1 = n² + n + 1.

Dual arguments show that π contains n² + n + 1 lines.

Section 5: The art of counting; results of Catalan, Euler and Gregory

Problem 92

(a) The number R_{n+1} of paired products of n + 1 factors $f_1, f_2, \ldots, f_{n+1}$ can be derived from the number R_n of paired products of n factors as follows.

An arbitrary paired product P of R_n consists of $n - 1$ paired multiplications of the form $A \cdot B$ (for example, the paired product of four factors: $[f_1 \cdot (f_2 \cdot f_3)] \cdot f_4$ consists of three paired multiplications: $f_2 \cdot f_3, f_1 \cdot (f_2 \cdot f_3)$ and $[f_1 \cdot (f_2 \cdot f_3)] \cdot f_4$).

P can be extended to a product P' of R_{n+1} by adjoining f_{n+1} to $A \cdot B$ in four different ways:

$$(f_{n+1} \cdot A) \cdot B, (A \cdot f_{n+1}) \cdot B, A \cdot (f_{n+1} \cdot B) \text{ and } A \cdot (B \cdot f_{n+1}).$$

For the $n - 1$ paired multiplications in P this leads to $(n - 1) \cdot 4$ different products P'. Moreover, P' can be obtained from P also by forming the products

$$f_{n+1} \cdot P \text{ or } P \cdot f_{N+1}.$$

Thus there are altogether $(n - 1) \cdot 4 + 2 = 4n - 2$ products P' of $n + 1$ factors obtained from a given product P of $f_1, f_2, \ldots, f_n$. This implies that

$$R_{n+1} = (4n - 2)R_n.$$

Since $R_1 = 1$, formula (38) leads to the following expression for R_n:

$$R_n = 2 \cdot 6 \cdot 10 \cdots \cdots (4n - 6).$$

(b) Our next task is to determine C_n, that is the number of those paired products of n factors $f_1, f_2, \ldots, f_n$ in which the order of the factors is prescribed.

Suppose that in such an arbitrary product p the brackets are left unaltered while the factors are permuted. The resulting new product P is one of the R_n products investigated in (a). Since the factors can be permuted in $n!$ ways, this implies that p corresponds to $n!$ paired products of type (a). Thus

$$C_n = \frac{R_n}{n!} = \frac{1 \cdot 2 \cdot 6 \cdots \cdots (4n - 6)}{n!}.$$

Problem 93

Let P_n be an arbitrary paired product of n factors $f_1, f_2, \ldots, f_n$ taken in a prescribed order. The last step in forming P_n consists of a paired multiplication of the form $P_n = P_i \cdot P_{n-i}$, where P_i is a paired product of the factors $f_1, f_2, \ldots, f_i$, and P_{n-i} is a paired product of the remaing factors $f_{i+1}, f_{i+2}, \ldots, f_n$. For a fixed i the number of products P_i is C_i and the number of products P_{n-i} is C_{n-i}. Since in forming $P_n = P_i \cdot P_{n-i}$ the number i can take any value from 1 to $n - 1$ inclusive, C_n satisfies the recurrence relation:

$$C_n = C_1 \cdot C_{n-1} + C_2 \cdot C_{n-2} + \cdots + C_{n-1} \cdot C_1.$$

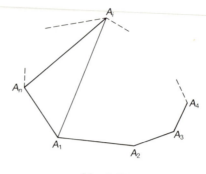

Fig. 3.17

Problem 94

(a) Proof of (8): Let $P_n = A_1 A_2 A_3 \cdots A_n$ be a convex polygon, and let D_i be a division of P_n into triangles such that the side $A_1 A_n$ belongs to a triangle $A_1 A_n A_i$ of the division for some i.

The diagonals $A_1 A_i$ and $A_n A_i$ divide P_n into three regions: the triangle $A_1 A_n A_i$, the polygon $P_i = A_1 A_2 \cdots A_i$ and the polygon $P_{n+1-i} = A_i A_{i+1} \cdots A_n$. The number of the different divisions D_i for a fixed i is the product of the number E_i of divisions of P_i into triangles by the number E_{n+1-i} of divisions of P_{n+1-i} into triangles. Since i can take any value from 2 to $n - 1$ inclusive, it follows that the number of divisions of P_n into triangles by diagonals is

$$E_n = E_2 E_{n-1} + E_3 E_{n-2} + \cdots + E_{n-1} E_2.$$

(b) Proof of (9): Applying the recurrence formulae

$$E_{n+1} = E_2 E_n + E_3 E_{n-1} + \cdots + E_n E_2$$

and

$$C_n = C_1 C_{n-1} + C_2 C_{n-2} + \cdots + C_{n-1} C_1$$

the relation $E_n = C_{n-1}$ can be proved by induction as follows:

It is true that $E_2 = 1 = C_1$ and $E_3 = 1 = C_2$. Assume that $E_i = C_{i-1}$ for all $i = 2, 3, \ldots, n$. This implies that

$$E_{n+1} = C_1 C_{n-1} + C_2 C_{n-2} + \cdots + C_{n-1} C_1 = C_n.$$

Thus

$$E_n = C_{n-1} \text{ for all natural numbers } n > 2.$$

Problem 95

A zigzag permutation $p = a_1 a_2 a_3 \cdots a_n$ can be represented by a zigzag line z joining the points $P_1(1, a_1), P_2(2, a_2), \ldots, P_n(n, a_n)$ in a Cartesian coordinate system, as shown in Fig. 3.18.

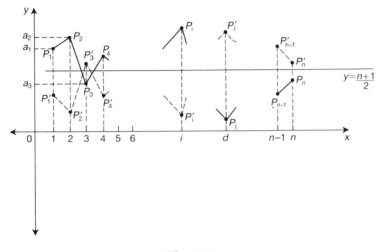

Fig. 3.18

Suppose that P_i is the highest point of z (with y-coordinate equal to n) and that P_j is the lowest point of z (with y-coordinate 1). Reflect z in the line with equation $y = (n + 1)/2$. The image $z' = P_1' P_2' \cdots P_n'$ of z corresponds to the permutation $p' = a_1' a_2' \cdots a_n'$, where $a_1', a_2', \ldots, a_n'$ are the y-coordinates of $P_1', P_2', \ldots, P_n'$ respectively. p' is also a zigzag permutation.

From the construction of p' it follows that if the permutation p starts by rising, then p' starts by falling and vice versa. Similarly, if p ends by rising or falling, then p' ends by falling or rising respectively.

Thus

$$A_n = B_n = \frac{Z_n}{2} \quad \text{and} \quad C_n = D_n = \frac{Z_n}{2}.$$

Problem 96

Let us first determine the number $Z_{n,i}$ of those zigzag permutations of $\{1, 2, \ldots, n\}$ in which the number n stands in the $(i + 1)$st place for some fixed i. For any such permutation p:

- the i places preceding n are filled with i numbers chosen arbitrarily from the set $\{1, 2, \ldots, n\}$ and arranged into zigzag permutation $a_1 a_2 \cdots a_i$. The permutation $a_1 a_2 \cdots a_i$ ends by falling ($a_i < a_{i+1} = n$);

- the remaining $n-1-i$ numbers from $\{1, 2, \ldots, n\}$, following n, form a zigzag permutation $a_{i+2}a_{i+3} \cdots a_n$, which starts by rising ($n = a_{i+1} > a_{i+2} < a_{i+3}$).

For any fixed choice of i numbers from the set $\{1, 2, \ldots, n-1\}$ there are D_i zigzag permutations $a_1 a_2 \ldots a_i$ which end by falling. To any such permutation there correspond A_{n-1-i} zigzag permutations $a_{i+2}a_{i+3} \cdots a_n$ which start by rising. Since the number of choices of i elements from a set of $n-1$ element is $\binom{n-1}{i}$, it follows that

$$Z_{n,i} = \binom{n-1}{i} D_i A_{n-1-i}.$$

The number n can occupy any of the n places in a zigzag permutation $a_1 a_2 \cdots a_n$; thus i can vary from 0 to $n-1$. This implies that the total number Z_n of zigzag permutations on $\{1, 2, \ldots, n\}$ is

$$Z_n = \sum_{i=0}^{n-1} Z_{n,i} = \sum_{i=0}^{n-1} \binom{n-1}{i} D_i A_{n-1-i},$$

It has been proved (see Problem 95) that $D_i = A_i$ and $Z_n = 2A_n$; hence

$$2A_n = \sum_{i=0}^{n-1} \binom{n-1}{i} A_i A_{n-1-i}, \tag{39}$$

Problem 97

$$\binom{n-1}{i} = \frac{(n-1)!}{i!(n-1-i)!},$$

and by definition $p_k = A_k/k!$. Hence (39) can be rewritten in the form

$$2p_n n! = \sum_{i=0}^{n-1} (n-1)! \, p_i p_{n-1-i},$$

or

$$2np_n = \sum_{i=0}^{n-1} p_i p_{n-1-i}. \tag{40}$$

Problem 98

(a) For $n \geqslant 3$ the number Z_n of zigzag permutations of the set $\{1, 2, \ldots, n\}$ is smaller than the number $n!$ of all possible permutations of this set. That is

$$Z_n < n!$$

or

$$2A_n = 2n!p_n < n!.$$

Hence

$$p_n < \tfrac{1}{2} \text{ for } n \geqslant 3.$$

(b) Since $p < \tfrac{1}{2}$ if $n \geqslant 3$, for $|x| < 1$ the function

$$y = p_0 + p_1 x + p_2 x^2 + \cdots$$

can be differentiated term by term (see [62]), and

$$y' = p_1 + 2p_2 x + 3p_3 x^2 + \cdots.$$

Moreover, y can be squared ([62]):

$$y^2 = \sum_{n=1}^{\infty} b_n x^{n-1}, \qquad \text{where} \qquad b_n = \sum_{i=0}^{n-1} p_i p_{n-1-i} \quad \text{for } n \geq 2$$

$$\text{and } b_1 = 1.$$

Formula (40) implies that

$$y^2 = 1 + 2 \cdot 2p_2 x + 2 \cdot 3p_3 x^2 + 2 \cdot 4p_4 x^3 + \cdots,$$

that is

$$y^2 + 1 = 2y'.$$

Problem 99

The differential equation $y^2 + 1 = 2y'$ can be solved by separating the variables:

$$\frac{dx}{2} = \frac{dy}{y^2 + 1},$$

that is

$$C + \frac{x}{2} = \text{arc tan } y.$$

To find the value of the constant C, put $x = 0$ in $y = p_0 + p_1 x + p_2 x^2 + \cdots$. The corresponding value of y is $p_0 = 1$. Thus $C = \pi/4$ and

$$y = \tan\left(\frac{x}{2} + \frac{\pi}{4}\right).$$

Problem 100

The equality

$$\tan\left(\frac{x}{2} + \frac{\pi}{4}\right) = p_0 + p_1 x + p_2 x^2 + \cdots \tag{41}$$

holds for every x such that $|x| < 1$ (see [62]).

If x is replaced by $-x$, equation (41) is transformed into

$$\tan\left(\frac{\pi}{4} - \frac{x}{2}\right) = p_0 - p_1 x + p_2 x^2 - p_3 x^5 + \cdots. \tag{42}$$

The sum of $\tan\left(\frac{x}{2} + \frac{\pi}{4}\right)$ and $\tan\left(\frac{\pi}{4} - \frac{x}{2}\right)$ is equal to 2 sec x:

$$\tan\left(\frac{x}{2} + \frac{\pi}{4}\right) + \tan\left(\frac{\pi}{4} - \frac{x}{2}\right) = \frac{\tan(x/2) + 1}{1 - \tan(x/2)} + \frac{1 - \tan(x/2)}{1 + \tan(x/2)}$$

$$= \frac{(1 + \tan(x/2))^2 + (1 - \tan(x/2))^2}{1 - \tan^2(x/2)}$$

$$= 2\frac{1 + \tan^2(x/2)}{1 - \tan^2(x/2)}$$

$$= 2 \cdot \frac{1}{\cos^2(x/2) - \sin^2(x/2)}$$

$$= 2 \cdot \frac{1}{\cos x} = 2 \sec x. \tag{43}$$

On the other hand, (41) and (42) imply that

$$\tan\left(\frac{x}{2} + \frac{\pi}{4}\right) + \tan\left(\frac{\pi}{4} - \frac{x}{2}\right) = 2(p_0 + p_2 x^2 + p_4 x^4 + \cdots). \tag{44}$$

(43) and (44) yield the infinite series for sec x:

$$\sec x = p_0 + p_2 x^2 + p_4 x^4 + \cdots = \tfrac{1}{2}\left(Z_0 + Z_2 \frac{x^2}{2!} + Z_4 \frac{x^4}{4!} + \cdots\right).$$

Similarly, it can be verified that

$$\tan\left(\frac{x}{2} + \frac{\pi}{4}\right) - \tan\left(\frac{\pi}{4} - \frac{x}{2}\right) = 2 \tan x, \tag{45}$$

and that, at the same time,

$$\tan\left(\frac{x}{2} + \frac{\pi}{4}\right) - \tan\left(\frac{\pi}{4} - \frac{x}{2}\right) = 2(p_1 x + p_3 x^3 + \cdots). \tag{46}$$

Hence, according to (45) and (46):

$$\tan x = \tfrac{1}{2}\left(Z_1 + \frac{Z_3}{3!} x^3 + \frac{Z_5}{5!} x^5 + \cdots\right).$$

The trigonometric functions $y = \sec x$ and $y = \tan x$ can be expanded in the above series over any interval $(-h, h)$ such that $|h| < 1$. (In fact, by methods of complex analysis one can show that these series converge for every x such that $|x| < \pi/2$.)

IV A selection of elementary problems treated by eminent twentieth-century mathematicians

Introduction

What kinds of research work are professional mathematicians concerned with nowadays? Answers to this question are beyond the scope of our book. Nevertheless, we shall be able to discuss here a variety of elementary, beautiful and intriguing problems which have caught the attention of eminent contemporary scholars. In addition to being intrinsically interesting, the problems chosen for this chapter are connected with important branches of modern mathematics.

The geometrical problem of Sylvester–Gallai, described in Section 1, leads to a generalization in the theory of block designs (a field of mathematics with various applications, e.g. in coding theory). The competition problems described in Section 2 are related to deep questions in combinatorics of Ramsey numbers. Section 3 contains problems on lattice points; the latter have been used since Minkowski in number theory and are encountered in a wide range of mathematical topics. Finally, Section 4 is devoted to special cases and generalizations of Fermat's last theorem, a statement which has puzzled generations of mathematicians from the seventeenth century until the present day.

The problems in Chapter IV are recommended to advanced readers. For additional information about the topics treated in this chapter see Appendices I and III.

Part I: Problems

Section 1: The problem of Sylvester–Gallai and related questions in Euclidean geometry and in combinatorics

1.1 The problem of Sylvester–Gallai

In 1893 Sylvester submitted the following problem to the journal *Educational Times*.

Problem A

Let $\mathscr{P}$ be a set of finitely many points in the plane, not all on a common straight line. Is it true that there exists at least one straight line in the plane carrying exactly two points of $\mathscr{P}$?

No solution to the above question appeared in the journal and it is not known whether Sylvester could answer it. Forty years later, however, Erdös, unaware of Sylvester's question, formulated the same problem and conjectured that the answer to it is affirmative. Erdös' conjecture was first verified by Gallai in 1933.

In 1943 Erdös published Problem *A* for solution in the journal *American Mathematical Monthly*. This time several solutions were received; among them, Kelly's solution was considered by Erdös as the most ingenious.

Our present aim is to outline Kelly's solution.

Problem 101

Let $\mathscr{P}$ be a finite set of non-collinear points in the plane. Denote by $\mathscr{L}$ the set of straight lines carrying at least two points of $\mathscr{P}$. Consider the distances of all points of $\mathscr{P}$ from all lines of $\mathscr{L}$ and denote the smallest of these distances by d.

Prove that if $P \in \mathscr{P}$ and $\ell \in \mathscr{L}$ is a point–line pair such that their mutual distance is d, then ℓ carries exactly two points of $\mathscr{P}$.

1.2 Two generalizations of Sylvester–Gallai's problem in Euclidean geometry

The first generalization of Sylvester–Gallai's problem, treated here, concerns points and planes in three-dimensional space. The following result is due to Motzkin.

Problem 102

Let $\mathscr{P}$ be a finite set of points in three-dimensional space such that not all of them belong to a common plane.

(a) Prove that there exists a plane π in the space such that the points of $\mathscr{P}$ on π are situated on not more than two straight lines of π.

(b) Show, by constructing a counter-example, that the following statement is, in general, false: There exists a plane in space containing exactly three non-collinear points of $\mathscr{P}$.

The second generalization, pointed out by Erdös, is obtained by replacing the straight lines in Problem A by circles.

Problem 103

Let $\mathscr{P}$ be a finite set of points in a plane, no three collinear, and not all of them on a common circle. Prove that there is at least one circle in the plane carrying exactly three points of $\mathscr{P}$.

1.3 The number of lines in $\mathscr{L}$, and an intriguing discovery when generalization breaks down

The following statement can be deduced from Gallai's theorem.

Problem 104

Let $\mathscr{P}$ be a set of n points in the plane, not all of them collinear, and let $\mathscr{L}$ be the set of the straight lines carrying at least two points of $\mathscr{P}$. Prove that the number of lines in $\mathscr{L}$ is at least n.

If a set $\mathscr{P}$ of n points is such that all but one of its elements are on a common line, then the number of lines in $\mathscr{L}$ is also n (see Fig. 4.1). This shows that the inequality $|\mathscr{P}| \leq |\mathscr{L}|$, estimated in Problem 104, is the best possible.

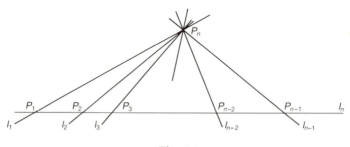

Fig. 4.1

Erdös conjectured that a similar result holds for the number of circles carrying at least three points of a non-concyclic finite point set in the plane. If all points but one of a set of n points belong to a common circle, then there are

$$\binom{n-1}{2} + 1 = \frac{(n-1)(n-2)}{2} + 1$$

circles carrying three or more points of the set (prove it); hence Erdös attempted to prove the following statement:

(S) If $\mathscr{P}$ is a set of n non-concyclic points in the plane, no three of which are collinear, and $\mathbb{C}$ is the set of circles carrying at least three points of $\mathscr{P}$, then $\mathbb{C}$ contains at least $\binom{n-1}{2} + 1$ elements.

Segre pointed out to Erdös that statement (S) is false by constructing the following counter-example.

Problem 105

Let $\mathbb{C}$ be a cube, $\mathscr{P}'$ the set of its vertices and $\mathbb{C}'$ the set of circles carrying at least three points of $\mathscr{P}'$.
   Construct a stereographic projection $\mathscr{P}$ of $\mathscr{P}'$ and $\mathbb{C}$ of $\mathbb{C}'$ onto a plane such that $|\mathbb{C}| \leq \binom{|\mathscr{P}| - 1}{2} + 1$.

1.4 A generalization of Gallai's result in the theory of block designs

The theory of block designs originated in statistics. It is an important branch of modern combinatorics. There are different types of block designs. In this section we shall define a block design $\mathscr{D}$ as follows:

$\mathscr{D}$ consists of n elements, called *points*, and of certain subsets of points, called *blocks*, such that:

- each block contains at least two distinct points; and
- any two distinct points are contained in exactly one block.

   Examples of block designs are finite projective planes (see Chapter III, Section 4.2). $\mathscr{P}$ and $\mathscr{L}$ in the Sylvester–Gallai problem (see Problem 101) are the point- and block-sets respectively of a block design, where different blocks do not necessarily contain the same number of points.
   In 1938 Hanani proved the following generalization of Gallai's result for block designs $\mathscr{D}$.

Theorem 11 If a block design $\mathscr{D}$ consists of n points and m blocks such that $m > 1$, then $m \geq n$.

   A simple, elegant proof of Theorem 11, due to de Bruijn, is described in Part II of this chapter (see Solution to Problem 106):

Problem 106

Prove Theorem 11.

Section 2: The pigeon-hole principle and some Ramsey numbers

2.1 A Hungarian competition problem and its generalization

If n objects are placed in k boxes, where $n = qk + r$, q and r are integers, and $0 < r < k$, then at least one box must contain more than q objects.

The above statement, whose truth is obvious, is one of the most powerful tools of combinatorics with applications in various branches of mathematics. It is known as *Dirichlet's principle* because Dirichlet (nineteenth century) applied it to approximate irrational numbers by rationals (see [60]). Dirichlet's principle is often called the *pigeon-hole principle*.

A straightforward application of the pigeon-hole principle leads to the solution of the following problem, posed in the Hungarian Mathematical Olympaid in 1947:

Problem 107

Prove that among six people in a room there are at least three who know one another, or at least three who do not know one another.

Problem 107 is closely related to Problem 108:

Problem 108 (VIth IMO, 1964)

Seventeen scientists correspond with one another. The correspondence is about three topics; any two scientists write to one another about one topic only.
Prove that at least three scientists write to one another on the same topic.

Problems 107 and 108 can be generalized as follows:

Problem 109 (*American Math. Monthly*, 1964, E 1653)

Let $\mathcal{P}_n$ be a set of $[en!] + 1$ points in the plane.[1] Any two dinstinct points of $\mathcal{P}_n$ are joined by a straight line segment, and each segment is coloured in one of n given colours. Show that among the segments there are at least three of the same colour, forming a triangle.

2.2 Some Ramsey numbers

Problem 109 raises the following question:

(Q)   What is the smallest possible number m_n of elements in a point set S_n

---

[1] $[en!]$ is the largest integer not greater than $en!$ and $e = 1/0! + 1/1! + 1/2! + 1/3! + \cdots$.

such that among the straight line segments, joining the point pairs in S_n, and coloured with n colours in an arbitrary way, there is a triple of the same colour forming a triangle?

For $n = 2$ and $n = 3$ one can show that $m_n = [en!] + 1$, that is $m_2 = 6$ and $m_3 = 17$ (see Problems 110 and 111).

Erdös conjectured that $m_n = [en!] + 1$ for all natural numbers n but so far the conjecture could not be verified. In its general form question (Q) remains open. We shall consider the cases $n = 2$ and $n = 3$.

Problem 110

Prove that $m_2 = 6$ by constructing a set S of five points such that it is possible to colour the straight line segments, joining the point pairs of S using two colours so that the configuration does not contain a triangle consisting of line segments of the same colour.

Problem 111

Verify that $m_3 = 17$ by constructing the following counter-example [94].

Let $G(\circ)$ be an elementary abelian group[1] of order 16 with generators a, b, c, d. Divide the elements of G, different from the neutral element O, into three disjoint subsets:

$$S_B = \{a, b, c, d, a \circ b \circ c \circ d\},$$
$$S_R = \{a \circ b, a \circ c, c \circ d, a \circ b \circ c, b \circ c \circ d\},$$
$$S_G = \{b \circ c, a \circ d, b \circ d, a \circ c \circ d, a \circ b \circ d\}.$$

Construct a coloured graph Γ as follows: The vertices of Γ are the elements of G. Any two distinct vertices of Γ are joined by a unique edge. The edge joining x and y is coloured blue, red or green if the element $x \circ y$ of G belongs to S_B, S_R or S_G respectively.

Prove that Γ contains no triangle with edges of the same colour.

m_2 and m_3 are examples of the so-called Ramsey numbers, and the statement of Problem 109 is a special case of a celebrated theorem due to Ramsey (twentieth century).

Theorem 12 (Ramsey's theorem)[2]   Let S be a set of N elements and let $r, t, q_1, q_2, \ldots, q_t$ be natural numbers, with $N \geq q_i \geq r$ for $i = 1, 2, \ldots, t$. Let T be the set of r-subsets (that is, subsets of r elements) of S.

Suppose that T is divided into t mutually disjoint subsets $T_1, T_2, \ldots, T_t$. Then there exists a least positive integer $R(q_1, q_2, \ldots, q_t, r)$, depending only on $q_1, q_2, \ldots, q_t, r$ and not on the set S, such that if $N \geq R(q_1, q_2, \ldots, q_t, 2)$, then for at least one $i \in \{1, 2, \ldots, t\}$ there is a q_i-subset S_i of S all of whose r-subsets are in T_i.

[1] See Appendix I.
[2] For a proof of Ramsey's theorem see e.g. [93].

Ramsey's result is a so-called *existence theorem*. It simply asserts that the number $R(q_1, q_2, \ldots, q_t, r)$ exists for any given integers $q_i \geq r \geq 1, i = 1, 2, \ldots, t$. So far, the numbers $R(q_1, q_2, \ldots, q_t, r)$, known as *Ramsey numbers*, have been determined in few cases.

Problem 112

From the results proved in this section deduce that

$$R(3, 3, \ldots, 3, 2) \leq [et!] + 1$$
$$\underbrace{}_{t \text{ times}}$$

and that

$$R(3, 3, 2) = 6 \quad \text{and} \quad R(3, 3, 3, 2) = 17.$$

Section 3: Problems on lattice points

3.1 Lattice points and circles

A lattice point in a two-, three- or higher-dimensional Cartesian coordinate system is a point whose coordinates are integers. Lattice points play an important rôle in various branches of Mathematics. In the last century Minkowski, the founder of geometric number theory, proved a variety of statements on integers, using lattice points.

In 1957 Steinhaus posed the following question:

(Q) Is it possible to construct a circle in a two-dimensional Cartesian coordinate system such that its interior contains exactly n lattice points, for any natural number n?

Sierpiński [92] answered Steinhaus' question affirmatively, by proving the following:

Theorem 13 The point P $(\sqrt{2}, \frac{1}{3})$ has different distances from all lattice points in the plane.

Problem 113

(a) Prove Theorem 13.
(b) Deduce that the answer to question (Q) is 'yes'.

3.2 Schoenberg's generalization of Steinhaus' problem

Schoenberg wondered: How did Sierpiński find the point P $(\sqrt{2}, \frac{1}{3})$ in Theorem 13? What is the set of all points in the plane, each at different distances from all lattice points?

The above questions led Schoenberg to the following study. Instead of points with integer coordinates in the plane, Schoenberg considered points with rational coordinates, and extended the investigation to n-dimensional real space R^n for any natural number n.

Let us call points whose coordinates are rational numbers rational points. In this paragraph we shall describe Schoenberg's results on rational points in R^n for $n = 1$ and $n = 2$.

The case $n = 1$ concerns the real number line.

Problem 114

Find the set S_1 of all points on the real number line R^1, each of which has different distances from all rational points of R^1.

In the case $n = 2$, concerning the real plane, Schoenberg proved the following.

Theorem 14   The set S_2 of points in the two-dimensional Cartesian coordinate system having different distances from all rational points consists of those points $P(x_1, x_2)$ in the plane which do not belong to any line with equation

$$a_1 x_1 + a_2 x_2 + a_3 = 0,$$

where a_1, a_2, a_3 are rational numbers not all equal to zero.

Problem 115

Prove Theorem 14.

It is not difficult to generalize Theorem 14 to higher dimensional spaces. This is left to the reader.

Section 4: Fermat's last theorem and related problems

In 1637 Fermat made a famous remark in the margin of his copy of Bachet's edition of Diophantos' book:

(F)   The equation $x^n + y^n = z^n$ cannot be solved in natural numbers x, y, z for any given integer $n \geq 3$.

Further, Fermat claimed that he had found a truly wonderful proof of the above statement, but the margin was too small to write it down.

Since then, Fermat's statement, known as *Fermat's last theorem*, has been investigated by many celebrated mathematicians. Efforts to prove it have led to significant developments in number theory and modern algebra, but whether or not the statement is correct remains an open question. Partial answers have been found: for example, (F) was proved for $n = 3$ by Euler and Legendre (eighteenth century); the proof of (F) for $n = 4$ was given by Fermat himself. Recently, Faltings (1983) showed that for any $n > 4$ the equation $x^n + y^n = z^n$ has at most finitely many solutions in natural numbers x, y, z such that x, y, z are pairwise relatively prime.

Below we shall study the equation $x^n + y^n = z^n$ in some special cases.

Problem 116 [21]

Prove that if x, y, z and n are natural numbers such that $n \geq z$ then the relation

$$x^n + y^n = z^n$$

does not hold.

Problem 117 [21]

If n is an odd integer greater than 1, and x, y, z are natural numbers forming an arithmetic progression, then the relation $x^n + y^n = z^n$ does not hold.

Finally, consider the related question:

Problem 118 [95]

(a) Let p, q, r be natural numbers greater than 1 and pairwise relatively prime. Does the equation

$$x_1^p + x_2^q = x_3^r$$

have solutions in natural numbers x_1, x_2 and x_3?

(b) Let p and q be relatively prime integers greater than 1. Does the equation

$$x_1^p + x_2^p = x_3^q$$

have solutions in natural numbers x_1, x_2 and x_3?

(c) Generalize (a) and (b) to $k + 1$ variables $x_1, x_2, \ldots, x_{k+1}$.

Part II: Solutions

Section 1: The problem of Sylvester and Gallai, and related questions

Problem 101

Let $P \in \mathscr{P}$ and let $\ell \in \mathscr{L}$ be a point–line pair at the smallest possible mutual distance d. Let P' be the foot of the perpendicular from P to ℓ. Since $\ell \in \mathscr{L}$,

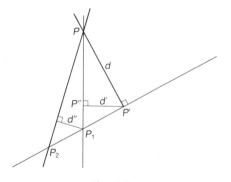

Fig. 4.2

there are at least two points P_1, P_2 of $\mathscr{P}$ on ℓ. One of them, say P_1, must be different from P'.

First we shall show that $P' \notin \mathscr{P}$. Otherwise, suppose that $P' \in \mathscr{P}$. The line PP_1, through two distinct points P and P_1 of $\mathscr{P}$, belongs to $\mathscr{L}$. Denote the foot of the perpendicular from P' to PP_1 by P'' and the distance $\overline{P'P''}$ of P' from PP_1 by d'. From the right-angled triangle $PP'P''$ it follows that

$$d' = \overline{P'P''} < \overline{PP'} = d.$$

Thus the distance of P' from PP_1 is smaller than the smallest possible distance between a point of $\mathscr{P}$ and a line of ℓ. This contradiction shows that the assumption $P' \in \mathscr{P}$ was false.

The next step is to show that ℓ cannot contain three distinct points of $\mathscr{P}$. Otherwise ℓ would contain at least two points of $\mathscr{P}$ on the same side of P'. Suppose that P_2 is a point of $\mathscr{P}$ on the same side of P' as P_1. Without loss of generality we can assume that P_1 is between P' and P_2. Denote the distance of P_1 from the line $PP_2 \in \mathscr{L}$ by d''. From the relations

$$\sin \not{\angle} P_1 P_2 P = \frac{d''}{P_2 P_1} = \frac{d}{PP_2} \quad \text{and} \quad P_2 P_1 < P_2 P' < PP_2$$

it follows that $d'' < d$, again a contradiction.

Thus the line ℓ contains exactly two points of $\mathscr{P}$, one on each side of P'.

Problem 102

(a) Since the points of $\mathscr{P}$ are not coplanar, $\mathscr{P}$ must contain at least four points P_1, P_2, P_3 and P_4 not in a common plane. (This implies that any three of the points P_1, P_2, P_3, P_4 are not collinear.) Let π be the plane through P_1, P_2, and P_3. For any point $P \in \mathscr{P}$, different from P_4, denote the point in which the line P_4P intersects π (if any) by P'. Let $\mathscr{P}'$ be the set of all points P'. The

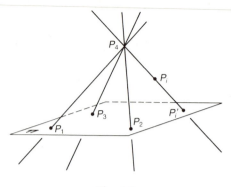

Fig. 4.3

set $\mathscr{P}'$ is not empty, since it contains $P'_1 = P_1$, $P'_2 = P_2$ and $P'_3 = P_3$. More-over, since P_1, P_2, and P_3 are not collinear, not all points of $\mathscr{P}'$ belong to a common line. According to the theorem of Gallai, there is a line ℓ' in π carrying exactly two points, say P'_i and P'_j of $\mathscr{P}'$. Hence, in the plane through P_4, P'_i and P'_j, all points which belong to $\mathscr{P}$ must lie on one (or both) of the lines $P_4P'_i$ and $P_4P'_j$.

(b) Figure 4.4 shows a pyramid with apex S and base $A_1B_1C_1$. The base $A_1B_1C_1$ is contained in a plane π_1. A plane π_2 intersects the edges SA_1, SB_1, SC_1 and the plane π_1 in the points A_2, B_2, C_2 and in the line ℓ, respectively. The lines A_2B_2 and A_1B_1 meet in a point P, the lines B_2C_2 and B_1C_1 in Q, and C_2A_2 and C_1A_1 in R. The points P, Q and R are on ℓ. It is easy to check that the set $\mathscr{P} = \{S, A_1, B_1, C_1, A_2, B_2, C_2, P, Q, R\}$ has the required property: there is no plane in the space carrying exactly three non-collinear points of $\mathscr{P}$.

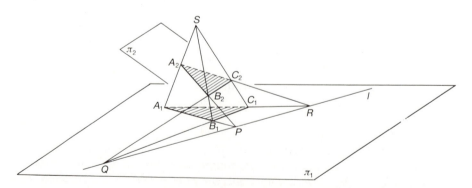

Fig. 4.4

Problem 103 can be solved easily by applying inversion (for properties of inversion see Appendix I)

Let P_1 be a point of $\mathscr{P}$ and let c be an arbitrary circle in the plane with centre P_1. Let α be the inversion with respect to c. Under α the points of $\mathscr{P}$ different from P_1 are mapped onto points that form a finite, non-collinear set $\mathscr{P}'$. According to Gallai's theorem there is a straight line ℓ in the plane carrying exactly two points, say P_2' and P_3', of $\mathscr{P}'$. Since no three points of $\mathscr{P}$ are collinear, ℓ does not pass through P_1. (Otherwise the image ℓ' of ℓ under α would be a straight line carrying P_1, P_2, and P_3.) Hence ℓ' is a circle through P_1; it carries exactly three points of $\mathscr{P}$: P_1, P_2, and P_3.

Problem 104

In view of Gallai's theorem there is a line $\ell_1 \in \mathscr{L}$ carrying exactly two points of $\mathscr{P}$. Remove one of these points, say P_1, from P, and remove from $\mathscr{L}$ all lines through P_1 containing exactly two points of $\mathscr{P}$. Denote the newly obtained set of points of $\mathscr{P}_1$ and the newly obtained set of lines by $\mathscr{L}_1$. The number $|\mathscr{L}_1|$ of lines in $\mathscr{L}_1$ is at most equal to $|\mathscr{L}| - 1$, that is

$$|\mathscr{L}_1| \leq |\mathscr{L}| - 1.$$

If the points of $\mathscr{P}_1$ are collinear, then in $\mathscr{L}$ there are exactly $n - 1$ lines connecting P_1 to points of $\mathscr{P}_1$. Hence $|\mathscr{L}| = (n - 1) + 1 = n$.

If the points of $\mathscr{P}_1$ are not collinear, then, according to Gallai's theorem, there is a line $\ell_1 \in \mathscr{L}_1$ containing exactly two points of $\mathscr{P}_1$. Remove one of these points, say P_2, from $\mathscr{P}_1$ and remove from $\mathscr{L}_1$ all lines through P_2 carrying exactly two points of $\mathscr{P}_1$. Call the newly obtained set of points $\mathscr{P}_2$ and the newly obtained set of lines $\mathscr{L}_2$. If the points of $\mathscr{P}_2$ are collinear then $|\mathscr{L}_1| = n - 1$ and $|\mathscr{L}| \geq |\mathscr{L}| + 1 = n$.

If the points of $\mathscr{P}_2$ are non-collinear, then construct, as above, $\mathscr{P}_3$ and $\mathscr{L}_3$. Continue the process of removing points and lines from $\mathscr{P}_3$ and $\mathscr{L}_3$ to construct $\mathscr{P}_4$, $\mathscr{L}_4$, $\mathscr{P}_5$, $\mathscr{L}_5$, . . . until a set $\mathscr{P}_k$ of collinear points is obtained. (This will happen in at most $n - 2$ steps.) In that case $|\mathscr{L}_{k-1}| = n - (k - 1)$. Since $|\mathscr{L}_{k-1}| \leq |\mathscr{L}| - (k - 1)$, it follows that

$$|\mathscr{L}| \geq |\mathscr{L}_{k-1}| + k - 1 = n - (k - 1) + (k - 1) = n.$$

Problem 105

$\mathscr{C}'$ consists of three types of circles:

1. circles circumscribing faces of $\mathbb{C}$ (e.g. the circle through A, B, C, and D);
2. circles through two diagonals of $\mathbb{C}$ (e.g. the circle through A, C, G, and E);
3. circles carrying exactly three vertices of $\mathbb{C}$ (e.g. the circle through A, F, and H).

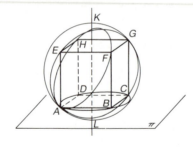

Fig. 4.5

There are 6 circles of type (1), 6 circles of type (2) and 8 circles of type (3). Thus there are altogether 20 circles in $\mathscr{C}'$.

Let S be the sphere circumscribed around S, and let KL be the diameter of S through the centres of the faces $ABCD$ and $EFGH$. Denote by π the plane touching S at L.

The points of $\mathscr{P}'$ and the circles of $\mathscr{C}'$ are on S. Construct their stereographic projection onto π from centre K. The projection of $\mathscr{P}'$ is a set $\mathscr{P}$ of 8 points. It is known that the stereographic projection of a circle on S not containing the centre of projection (K) is a circle on π (see Appendix I). Since K does not lie on any circle of $\mathscr{C}'$, the images of $\mathscr{C}'$ form a set $\mathscr{C}$ of 20 circles in π.

The circles of $\mathscr{C}$ are exactly those circles of π which carry at least three points of $\mathscr{P}$. The points of $\mathscr{P}$ are not concyclic. Since

$$\binom{|\mathscr{P}| - 1}{2} + 1 = \binom{7}{2} + 1 = 22,$$

the relation

$$|\mathscr{C}| \geq \binom{|\mathscr{P}| - 1}{2} + 1$$

is false.

Problem 106

Denote the points of $\mathscr{D}$ by $P_1, P_2, \ldots, P_n$ and the blocks by $b_1, b_2, \ldots, b_m$. Further, denote the number of blocks containing P_i by r_i for $r = 1, \ldots, n$, and the number of points in b_j by k_j for $j = 1, \ldots, m$. Since the number of blocks is greater than 1, it follows that $1 < k_j < n$. Also, $1 < r_i < m$.

The number I of the point-block pairs P_i, b_j such that P_i belongs to b_j can be counted in two ways:

1. by counting the blocks through the points, which gives

$$I = r_1 + r_2 + \cdots + r_n;$$

2. by counting the points on the blocks, which gives

$$I = k_1 + k_2 + \cdots + k_m.$$

From (1) and (2) it follows that

$$r_1 + r_2 + \cdots + r_n = k_1 + k_2 + \cdots + k_m. \tag{1}$$

We can assume, without loss of generality, that the points are labelled so that

$$r_j \leq r_i \qquad \text{if } j > i. \tag{2}$$

Moreover, the labelling of the blocks can be carried out so that the blocks passing through P_n are $b_1, b_2, \ldots, b_{r_n}$.

Our aim is to find further connections between the numbers k_j and r_i. For this note the following.

If P_i is any point and b_j any block not containing P_i, then through any point of b_j there passes exactly one block containing P_i. Moreover, blocks through P_i which pass through distinct points of b_j are different; this is so because any two points in $\mathcal{D}$ are contained in exactly one block. Thus

$$k_j \leq r_i. \tag{3}$$

Applying (3) to P_n and to each of the blocks $b_{r_n+1}, b_{r_n+2}, \ldots, b_m$, we see that

$$k_{r_n+i} \leq r_n \qquad \text{for } i = 1, 2, \ldots, m - r_n. \tag{4}$$

By adding the inequalities (4) together we get:

$$k_{r_n+1} + k_{r_n+2} + \cdots + k_m \leq (m - r_n)r_n. \tag{5}$$

$k_1, k_2, \ldots, k_{r_n}$ can be estimated as follows.

The block b_1 passes through at least one point, different from P_n, say through P_1. In that case $b_2, b_3, \ldots, b_{r_n}$ cannot pass through P_1. Hence, in view of (3),

$$r_1 \geq k_2.$$

The block b_2 also contains a point different from P_n (and from P_1), say P_2. The point P_2 is not on $b_3, \ldots, b_{r_n}$; thus

$$r_2 \geq k_3.$$

Continuing in this way, one obtains the inequalities

$$r_i \geq k_{i+1} \qquad \text{for } i = 1, \ldots, r_n - 1. \tag{6}$$

Finally, there is a point P_{r_n} on b_{r_n}, not contained in b_1; therefore

$$r_{r_n} \geq k_1. \tag{7}$$

Adding inequalities (6) and (7) together, we find that

$$r_1 + r_2 + \cdots + r_{r_n} \geq k_1 + k_2 + \cdots + k_{r_n}. \tag{8}$$

Combining (8) and (5) yields

$$r_1 + r_2 + \cdots + r_{r_n} + (m - r_n)r_n \geq k_1 + k_2 + \cdots + k_m. \tag{9}$$

(9), (1) and (2) imply that

$$r_1 + r_2 + \cdots + r_{r_n} + (m - r_n)r_n \geq r_1 + r_2 + \cdots + r_n$$

$$\geq r_1 + r_2 + \cdots + r_{r_n} + (n - r)r_n,$$

that is

$$m \geq n.$$

This proves Hanani's theorem.

Section 2: The pigeon-hole principle and some Ramsey numbers

Problem 107
Denote the six people by six points: A_1, A_2, A_3, A_4, A_5 and A_6. Join each pair of distinct points by a line segment and colour this segment red if the persons represented by the points know each other; otherwise colour the segment blue.

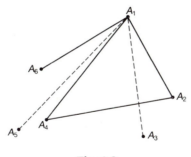

Fig. 4.6

Consider the line segments through one of the points, say A_1. Since there are five such line segments, at least three of them, e.g. A_1A_i, A_1A_j, A_1A_k must have the same colour (according to the pigeon-hole principle). Suppose that this colour is red. If any of the line segments A_iA_j, A_jA_k or A_kA_i is red, for example A_iA_j, then the three persons, corresponding to the points A_1, A_i, A_j know each other. Otherwise the line segments A_iA_j, A_jA_k and A_kA_i form a blue triangle. This means that the people represented by A_i, A_j and A_k do not know one another.

A similar conclusion will be reached if A_1A_i, A_1A_j, and A_1A_k are all blue.

Problem 108

Represent the 17 scientists by 17 points A_1, A_2, . . ., A_{17} and the three topics on which they correspond by three colours: red, blue, and green. Join any pair of points A_i, A_j by a line segment, and colour the segment using the colour of the topic on which the scientists represented by A_i and A_j correspond.

Consider the 16 line segments passing through A_{17}. Since $16 = 3 \times 5 + 1$, by the pigeon-hole principle there are at least 6 line segments of the same colour through A_{17}. Without loss of generality, we may suppose that A_1A_{17}, A_2A_{17}, A_3A_{17}, A_4A_{17}, A_5A_{17} and A_6A_{17} are all coloured green. We distinguish two cases:

Case 1:   One of the line segments joining the points A_1, A_2, A_3, A_4, A_5, A_6 is green, say A_iA_j, where $i, j \in \{1, 2, . . ., 6\}$, $i \neq j$. In that case all sides of the triangle $A_1A_iA_j$ are green. This means that the scientists labelled A_1, A_i, A_j write to one another on the same topic, represented by the green colour.

Case 2:   All line segments joining A_1, A_2, A_3, A_4, A_5, and A_6 are red or blue. This reduces Case 2 to Problem 107. According to Problem 107, among the triangles with vertices from the set $\{A_1, A_2, A_3, A_4, A_5, A_6\}$ there is at least one with sides of the same colour.

Problem 109

For $n = 2$ or $n = 3$ the number $[en!] + 1$ is 6 or 17, respectively. In these special cases the statement of Problem 109 is true, according to Problems 107 and 108. If $n = 1$ then $\{en!\} + 1 = 3$; in this trivial case the statement is also correct.

In the general case the idea for a solution is suggested by the method of solving Problem 108 with the help of Problem 107.

For any integer $k \geq 2$ denote by s_k the smallest number of points in a set S_k with the following property.

Among the line segments joining a point of S_k to the remaining points, and coloured in one of k given colours, there are at least s_{k-1} segments of the same colour.

According to the pigeon-hole principle, the above condition implies the following connection between s_k and s_{k-1}:

$$s_k - 1 = k(s_{k-1} - 1) + 1, \tag{10}$$

or, after dividing by $k!$,

$$\frac{s_k - 1}{k!} = \frac{s_{k-1} - 1}{(k-1)!} + \frac{1}{k!} \tag{11}$$

Putting $s_0 = 1$ and applying repeatedly the recurrence formula (11), one finds that

$$\frac{s_k - 1}{k!} = \frac{1}{0!} + \frac{1}{1!} + \frac{1}{2!} + \cdots + \frac{1}{k!}. \tag{12}$$

It is well known that

$$\frac{1}{0!} + \frac{1}{1!} + \frac{1}{2!} + \cdots + \frac{1}{k!} + \frac{1}{(k+1)!} + \frac{1}{(k+2)!} + \cdots = e.$$

Hence, according to (12),

$$e = \frac{s_k - 1}{k!} + r_k$$

where

$$r_k = \frac{1}{(k+1)!} + \frac{1}{(k+2)!} + \cdots < \frac{1}{k!}\left(\frac{1}{k+1} + \frac{1}{(k+1)^2} + \frac{1}{(k+1)^3} + \cdots\right)$$

$$= \frac{1}{k!} \cdot \frac{1}{k}.$$

This implies that

$$\frac{s_k - 1}{k!} \le e \le \frac{s_k - 1}{k!} + \frac{1}{k!} \cdot \frac{1}{k}.$$

or

$$s_k - 1 < ek! < s_k - 1 + \frac{1}{k}.$$

Thus

$$[ek!] = s_k - 1,$$

i.e.

$$s_k = [ek!] + 1.$$

This completes the solution of Problem 109.

Problem 110
Figure 4.7 shows a graph Γ. The vertices of Γ are the vertices of a convex pentagon; the edges of Γ are the sides and the diagonals of the pentagon. The sides of the pentagon are coloured red and the diagonals are coloured blue. Γ does not contain any triangle with edges of the same colour.

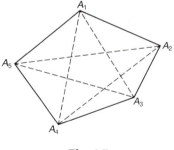

Fig. 4.7

Problem 111
In an elementary abelian group of order 16, for any element x it is known that $x \circ x = 0$ (see Appendix I). Using this property one can check easily that the 'sum' $x \circ y$ of any two elements x, y from S_B does not belong to S_B.

Similarly, the 'sum' of any two elements from S_R or S_G does not belong to S_R or S_G respectively.

Suppose that the graph Γ, described in Problem 111, contains three vertices x, y, z such that the edges joining these vertices are of the same colour. Thus, according to the rules for colouring Γ, the elements $x \circ y$, $x \circ z$ and $y \circ z$ of the group belong to a common subset S_i for some $i \in \{B, R, G\}$.

However, $y \circ z$ is the 'sum' of $x \circ y$ and $x \circ z$; this is so because

$$(x \circ y) \circ (x \circ z) = x \circ x \circ y \circ z = y \circ z.$$

Therefore $y \circ z$ cannot belong to the same subset S_i as $x \circ y$ and $x \circ z$. This contradiction shows that Γ does not contain a triangle with edges of the same colour.

Problem 112

In a set S of N elements consider all subsets of 2 elements (that is, $r = 2$). Represent the elements of S by points, and the 2-subsets of S by line segments joining the corresponding pairs of points. Denote by T the set of all 2-subsets of S and divide T into t mutually disjoint subsets $T_1, \ldots, T_t$. Colour the line segments corresponding to a point pair in T_i using a colour c_i so that $c_i \neq c_j$ whenever $i \neq j$. Thus a coloured graph Γ is obtained.

1. The theorem of Ramsey asserts that there exists a smallest natural number $R(\underbrace{3, 3, 3, \ldots, 3, 2}_{t \text{ times}})$ such that for $N \geq R(3, 3, 3, \ldots, 3, 2)$ there is a 3-subset S_i of S whose 2-subsets all belong to a common T_i for some $i = 1, 2, \ldots, t$.

    This implies that the three points representing the elements of S_i are joined by line segments of the same colour c_i, forming a triangle.
2. According to the solution of Problem 109, if Γ has at least $[et!] + 1$ nodes than it contains a monochromatic triangle.

(i) and (ii) imply that

$$R(\underbrace{3, 3, \ldots, 3, 2}_{t \text{ times}}) \leq [et!] + 1.$$

In the special cases, when $t = 2$ or $t = 3$, Problems 110 and 111 show that $R(3, 3, 2)$ and $R(3, 3, 3, 2)$ cannot be less than 6 or 17 respectively. Thus

$$R(3, 3, 2) = [e \cdot 2!] + 1 = 6$$

and

$$R(3, 3, 3, 2) = [e \cdot 3!] + 1 = 17.$$

Section 3: Problems on lattice points

Problem 113

(a) Let $A(a, b)$ and $B(c, d)$ be two different lattice points. In other words, a, b, c, d are integers and at least one of the relations $a \neq c$, $b \neq d$ is valid. Suppose that $PA = PB$, that is

$$\sqrt{[(\sqrt{2} - a)^2 + (\tfrac{1}{3} - b)^2]} = \sqrt{[(\sqrt{2} - c)^2 + (\tfrac{1}{3} - d)^2]},$$

leading to

$$a^2 + b^2 - c^2 - d^2 - \tfrac{2}{3}b + \tfrac{2}{3}d = 2\sqrt{2}(a - c).$$

Since a, b, c, d are integers, the last equation implies that

$$a - c = 0$$

and

$$b^2 - d^2 - \tfrac{2}{3}(b - d) = 0.$$

From $a = c$ it follows that $b \neq d$, that is $b - d \neq 0$. Hence $b + d = \tfrac{2}{3}$, which is impossible, b and d being integers.

Theorem 13 is proved.

(b) Let c_r be the circle of radius r with centre $P(\sqrt{2}, \tfrac{1}{3})$, and let $f(r)$ be the number of lattice points inside c_r. Then $f(r)$ has the following properties:

1. $f(r) = 0$ for small values of r (e.g. $f(0.1) = 0$).
2. $f(k + 1) > k^2$ for any natural number k. (The interior of c_{k+1} contains a square s_k with sides parallel to the coordinate axes and of side length $a_k > k$; hence the number of lattice points inside c_{k+1} is at least k^2.)
3. $f(r)$ increases by unit jumps, as r increases; this is so in view of (a).

From (1)–(3) it follows that $f(r)$ takes all positive integer values. Thus, for any natural number n there is a circle with centre $P(\sqrt{2}, \tfrac{1}{3})$ containing exactly n lattice points.

Problem 114

Denote by Q^1 the set of rational lattice points on the number line R^1, and by I^1 the set of irrational points on R^1.

Any rational point r is at equal distances from the rational points $r - r'$ and $r + r'$ for $r' \in Q^1$. Hence $r \in S^1$.

On the other hand, any irrational point i is at different distances from any two rational points r_1, r_2 on R_1. (Otherwise, $|r_1 - i| = |r_2 - i|$, $r_1 < r_2$ would imply that $r_2 - i = i - r_1$, that is $i = (r_1 + r_2)/2 \in Q^1$, which is false.)

Thus $S^1 = I^1$.

Problem 115

Denote by Q^2 the set of rational lattice points in a two-dimensional Cartesian coordinate system. Let I^2 be the set of points in R^2 each of which has different distances from all points of Q^2.

The following observations prove to be useful:

1. A point $P(x_1, x_2) \in R^2$, at equal distances from two distinct points $A(p_1, p_2)$, $B(q_1, q_2) \in Q^2$ is on the perpendicular bisector $\pi(A, B)$ of the line segment AB.

    The equation of $\pi(AB)$:

    $$\sqrt{[(x_1 - p_1)^2 + (x_2 - p_2)^2]} = \sqrt{[(x_1 - q_1)^2 + (x_2 - q_2)^2]}$$

    can be rewritten in the form

    $$(2q_1 - 2p_1)x_1 + (2q_2 - 2p_2)x_2 + p_1^2 - q_1^2 + p_2^2 - q_2^2 = 0 \quad (13)$$

2. Points which have different distances from all rational points are precisely those points not on any $\pi(AB)$, where $A, B \in Q^2$.
3. A line ℓ is a $\pi(A, B)$ if and only if its equation is of the form $a_1 x_1 + a_2 x_2 + a_3 = 0$, where a_1, a_2, a_3 are rational numbers, not all 0.

The proof of (3) consists of two steps:

Step 1:   If the line ℓ is a $\pi(AB)$ its equation is given by (13), which is of the required form.

Step 2:   Let ℓ be a line with equation $a_1 x_1 + a_2 x_2 + a_3 = 0$

In the equation $a_1 x_1 + a_2 x_2 + a_3 = 0$ the coefficients a_1, a_2 are not both 0. (Since $a_1 = a_2 = 0$ would imply that also $a_3 = 0$, and we are given that a_1, a_2, a_3 are not all 0.) Suppose without loss of generality that $a_1 \neq 0$. Then the point $M(-a_3/a_1, 0)$ is on the line ℓ, and the perpendicular to ℓ through M carries the rational points $A(-a_3/a_1 + a_1, a_2)$ and $B(-a_3/a_1 - a_1, -a_2)$ (Fig. 4.8). Hence the line ℓ is the perpendicular bisector $\pi(AB)$ of the line segment AB.

Theorem 14 is a consequence of (2) and (3).

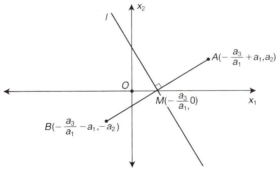

Fig. 4.8

Section 4: Fermat's last theorem and related problems

Problem 116

Suppose that x, y, z and n are natural numbers such that $n \geq z$ and $x^n + y^n = z^n$.

Clearly, $x < z$, $y < z$ and $x \neq y$. Without loss of generality, we may assume that $x < y$. Then

$$x^n = z^n - y^n = (z - y)(z^{n-1} + yz^{n-2} + \cdots + y^{n-1})$$
$$\geq 1 \cdot nx^{n-1} > x^n.$$

The above contradiction shows that the statement of Problem 116 is true.

Problem 117

Suppose that x, y, z are natural numbers satisfying the equation $x^n + y^n = z^n$ (where n is an odd integer greater than 1) and forming an arithmetic progression.

In this case z and x can be expressed in the form $z = y + d$, $x = y - d$, where d is a positive integer less than y. Equation $x^n + y^n = z^n$ can be rewriten as

$$(y - d)^n + y^n = (y + d)^n,$$

or, after dividing by d^n, as

$$\left(\frac{y}{d} - 1\right)^n + \left(\frac{y}{d}\right)^n = \left(\frac{y}{d} + 1\right)^n. \tag{14}$$

Set $y/d = t$ and expand the expressions in (14) to get

$$t^n - \binom{n}{1} t^{n-1} + \binom{n}{2} t^{n-2} - \cdots - 1 + t^n = t^n + \binom{n}{1} t^{n-1}$$
$$+ \binom{n}{2} t^{n-2} + \cdots + 1,$$

or

$$t^n - 2 \binom{n}{1} t^{n-1} - 2 \binom{n}{3} t^{n-3} - \cdots - 2 = 0. \tag{15}$$

(*Remark:*   The last summand on the left hand side of (15) is -2, since n is an odd integer.)

Our next task is to verify that equation (15) has no rational solution. This will be done in two steps:

Step 1:   Suppose that the rational number $t = r/s$, where r and s are relatively prime integers and $s > 1$ satisfies equation (15). In that case (15) can be rewritten as

$$r^n - s \left[2 \binom{n}{1} r^{n-1} + 2 \binom{n}{3} s^2 r^{n-3} + \cdots + 2s^{n-1} \right] = 0,$$

implying that r is divisible by s contradicting on initial assumption.

Step 2:   Suppose that $t = r$, an integer satisfying (15). t cannot be odd since, according to (15),

$$t^n = 2 \left[\binom{n}{1} t^{n-1} + \binom{n}{3} t^{n-3} + \cdots + 1 \right].$$

Similarly, t cannot be even, since in that case all summands in (15) except -2 would be divisible by 4. (Recall that $n > 1$.)

Since (15) is not satisfied by any rational number t, it follows that equation $x^n + y^n = z^n$ cannot be satisfied by three positive integers $y - d, y, y + d$ if n is odd and greater than 1.

Problem 118

(a) We shall tackle the problem in the general case. Our task is to find positive integer solutions $x_1, x_2, \ldots, x_{k+1}$ for the equation

$$x_1^{p_1} + x_2^{p_2} + \cdots + x_k^{p_k} = x_{k+1}^{p_{k+1}}, \tag{16}$$

where $k \geq 2$ and $p_1, p_2, \ldots, p_{k+1}$ are pairwise relatively prime natural numbers greater than 1.

We shall show that (16) has infinitely many solutions $x_1, x_2, \ldots, x_{k+1}$ of the form $x_i = k^{\alpha_i}$ for suitable natural numbers α_i, $i = 1, 2, \ldots, k+1$.

Suppose that α_i are positive integers such that

$$\alpha_i p_i = \lambda, \qquad i = 1, 2, \ldots, k,$$

for a certain, fixed natural number λ (which has to be determined). In that case, setting $x_i = k^{\alpha_i}$ for $i = 1, 2, \ldots, k+1$ reduces (16) to

$$k \cdot k^\lambda = k^{p_{k+1} \alpha_{k+1}}.$$

Hence

$$1 + \lambda = p_{k+1} \alpha_{k+1},$$

or

$$1 = p_{k+1} \alpha_{k+1} - \lambda. \tag{17}$$

Since λ is a multiple of the pairwise relatively prime numbers $p_1, p_2, \ldots, p_k$, it follows that λ is also a multiple of their product $P = p_1 p_2 \cdots p_k$. Let $\lambda = Pq$ for some positive integer q. Substituting Pq for λ in (17), we get

$$1 = p_{k+1} \alpha_{k+1} - Pq. \tag{18}$$

(18) represents a Diophantine equation with unknown quantities α_{k+1} and q. The coefficients p_{k+1} and P in the equation are relatively prime natural numbers.

It is well known (see e.g. [60]) that a Diophantine equation $1 = ax - by$, where a and b are relatively prime positive integers, has infinitely many solutions in natural numbers x, y. Moreover, if x_0, y_0 represent a solution of $1 = ax - by$, then all positive integer solutions of the equation are of the form

$$x = x_0 + bt, \qquad y = y_0 + at,$$

where t is an integer such that $t > \max\left(-\frac{x_0}{b}, -\frac{y_0}{a} \right)$,

Thus, we know that (18) has infinitely many positive integer solutions α_{k+1}, q. They are of the form

$$\alpha_{k+1} = (\alpha_{k+1})_0 + Pt, \qquad q = q_0 + p_{k+1}t, \tag{19}$$

where $(\alpha_{k+1})_0, q_0$ is a pair of solutions of (18) and t is an integer such that $t > \max(-(\alpha_{k+1})_0/P, -q_0/p_{k+1})$.

Since $\lambda = Pq = \alpha_i p_i$, and $q = q_0 + p_{k+1}t$, it follows that

$$\alpha_i = \frac{P(q_0 + p_{k+1}t)}{p_i} \qquad \text{for } i = 1, 2, \ldots, k.$$

We have shown that equation (16) has infinitely many solutions of the form

$$x_i = k^{\alpha_i}, \qquad i = 1, 2, \ldots, k+1.$$

(b) Again we shall consider the equation with $k + 1$ unknowns:

$$x_1^p + x_2^p + \cdots + x_k^p = x_{k+1}^q, \tag{20}$$

where p and q are relatively prime integers greater than 1.

We shall prove that equation (20) has infinitely many solutions of the form

$$x_i = k^\alpha \qquad \text{for } i = 1, 2, \ldots, k \qquad \text{and} \qquad x_{k+1} = k^\beta, \tag{21}$$

where α and β are natural numbers — which have to be determined.

By substituting (21) into (20), the latter is transformed into

$$k \cdot k^{\alpha p} = k^{\beta q}.$$

Hence

$$1 + \alpha p = \beta q. \tag{22}$$

It was pointed out in (a) that there are infinitely many pairs of positive integers α, β satisfying (22). They are of the form

$$\alpha = \alpha_0 + qt, \beta = \beta_0 + pt,$$

where α_0, β_0 represent a particular solution of (22); t is an integer greater than $\max(-\beta_0/p, -\alpha_0/q)$.

Thus equation (20) has infinitely many solutions of the form $x_1 = x_2 = \cdots = x_k = k^{\alpha_0 + qt}, x_{k+1} = k^{\beta_0 + pt}$.

Appendix I Definitions and basic results on the following topics:

(a) Numbers and number patterns.

(b) Polynomials and polynomial equations; diophantine equations.

(c) Examples of algebraic structures: groups and real vector spaces; vectors.

(d) Coordinate systems: Cartesian coordinate systems, barycentric coordinate systems and Argand diagrams.

(e) Convex point sets; polygons; inscribed and circumscribed circles.

(f) Geometric transformations (rotation about a point; rotation about an axis; inversion; stereographic projection; translation; reflection).

(g) Combinatorial concepts (permutations, combinations, binomial and trinomial coefficients; graphs; balanced incomplete block designs).

(h) Number sequences and limits.

(a) Numbers and number patterns

1. Definitions of numbers

- A *natural number* is a member of the set $\{1, 2, 3, 4, \ldots\}$.
- Natural numbers are also called *positive integers*, or *positive whole numbers*.
- An *integer* is a member of the set $\{\ldots, -3, -2, -1, 0, 1, 2, 3, \ldots\}$. The numbers, $-1, -2, -3, \ldots$ are called *negative integers*. 0 is an integer; it is neither positive nor negative.
- A *fraction* is a number of the form a/b, where a is any integer and b is any positive integer. Thus every integer is also a fraction (e.g. $-6 = -6/1$), but not every fraction is an integer (for example, $\frac{2}{3}$ is not an integer). Fractions are also called *rational numbers*.
- A *finite, non-negative decimal fraction* is a fraction p/q which can be expressed in the form

$$\frac{p}{q} = n_0 + \frac{n_1}{10} + \frac{n_2}{10^2} + \frac{n_3}{10^3} + \cdots + \frac{n_k}{10^k},$$

205

where n_0 is a non-negative integer and $n_i \in \{0, 1, 2, \ldots, 9\}$ for $i = 1, 2,$ $\ldots, k$. It is customary to write p/q in the form

$$p/q = n_0.n_1 n_2 \cdots n_k.$$

with the *decimal point* after n_0.

- A *finite, negative decimal fraction* is a negative fraction

$$\frac{p}{q} = -\left(n_0 + \frac{n_1}{10} + \frac{n_2}{10^2} + \cdots + \frac{n_k}{10^k}\right),$$

where n_0 is a non-negative integer, $n_i \in \{0, 1, 2, \ldots, 9\}$ for $i = 1, 2, \ldots,$ k and not all n_i, $i = 0, 1, 2, \ldots, k$ are equal to 0. In this case p/q is written in the form $p/q = -n_0.n_1 n_2 \cdots n_k$.

- The length $\sqrt{2}$ of the diagonal of a unit square (Fig. A.1) is not a rational number; it cannot be expressed as $\sqrt{2} = a/b$.

Fig. A.1

Instead, $\sqrt{2}$ can be enclosed between pairs of decimal fractions in a never-ending sequence:

$$1 < \sqrt{2} < 2$$
$$1.4 < \sqrt{2} < 1.5$$
$$1.41 < \sqrt{2} < 1.42$$
$$1.414 < \sqrt{2} < 1.415$$
$$1.4142 < \sqrt{2} < 1.143$$
$$\vdots$$

We say that $\sqrt{2}$ is represented by *an infinite decimal fraction* 1.414213562 $\ldots$ *with a non-recurring pattern of decimal digits*. It is not difficult to show that *infinite decimal fractions with recurring patterens of decimals*

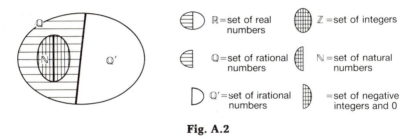

Fig. A.2

are rational numbers (e.g. $0.333\cdots = \frac{1}{3}$ or $2.035121212 = \frac{6109}{3000}$). On the other hand, infinite decimal numbers with no recurring patterns of decimals are not rational numbers. They are called *irrational numbers*.

- The set of *real numbers* is the union of the set of rational and of the set of irrational numbers. The Venn diagram in Fig. A.2 shows the set of real numbers together with the subsets of rational numbers, irrational numbers, integers and natural numbers

  The set $\mathbb{R}$ of real numbers has the following representation on the number line. The number line is a straight line ℓ (Fig. A.3) with a distinguished point O on ℓ, called the origin. An arbitrary real number r is represented by the point P_r on ℓ, which is on the same side of O as P_1 if $r > 0$, on the opposite side of O if $r < 0$, and, in either case, is at a distance $|r|$ from O. ($|r|$ is the *absolute value* of r. By definition $|r| = r$ if $r \geq 0$ and $|r| = -r$ if $r < 0$.)

- The *imaginary unit* is the number $\mathrm{i} = \sqrt{(-1)}$. The number i is not a real number.

- Real multiples of i, that is numbers of the form bi, where b is a real number, are called *imaginary numbers*.

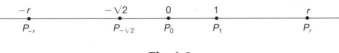

Fig. A.3

- Numbers of the form $a + ib$, where a and b are real numbers, are called *complex*. Thus real numbers may be thought of as complex numbers of the form $a + i0$. We denote the set of all complex numbers by $\mathbb{C}$.

  Addition and multiplication of complex numbers $z_1 = x_1 + iy_1$ and $z_2 = x_2 + iy_2$ are defined according to the rules:

$$z_1 + z_2 = (x_1 + x_2) + i(y_1 + y_2)$$

$$z_1 \cdot z_2 = (x_1 x_2 - y_1 y_2) + i(x_1 y_2 + x_2 y_1).$$

- For the representation of complex numbers on the Argand diagram see (d).
- *Quaternions* were invented by Hamilton as a generalization of complex numbers (see Chapter III, Section 3).

2. Factors, prime numbers, congruences

- A *factor* or a *divisor* of a natural number n is a natural number n_1 such that the quotient n/n_1 is also a natural number.
- A *prime number* is a natural number $p > 1$ whose only factors are 1 and p.
- A *composite number* is a natural number which is not a prime.
- Two natural numbers are *relatively prime or coprime* if their only common factor is 1.
- Let m be a non-negative integer. Two integers a and b are said to be *congruent modulo m* if their difference $a - b$ is a multiple of m. That is:

$$a \equiv b \ (\text{mod } m) \text{ if and only if } a - b = mt \text{ for some integer } t.$$

The expression

$$a \equiv b \ (\text{mod } m)$$

is called a *congruence*.

3. Figurate numbers

The notion of *figurate numbers* originated in ancient Greece. The Greeks represented special types of natural numbers by patterns of dots (or pebbles) in the shape of regular polygons and polyhedra. Such numbers are called figurate numbers. In particular:

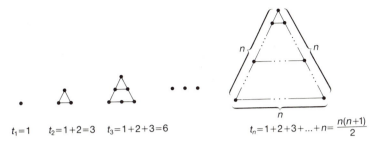

$t_1=1$ $t_2=1+2=3$ $t_3=1+2+3=6$ $t_n=1+2+3+...+n=\dfrac{n(n+1)}{2}$

Fig. A.4

- The *triangular numbers* $t_1, t_2, \ldots, t_n$, are the numbers of dots forming the equilateral triangles (Fig. A.4).
- The *square* numbers $s_1, s_2, \ldots, s_n$ are the numbers of dots forming squares (Fig. A.5).

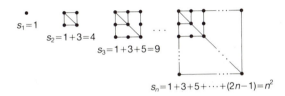

$s_1=1$

$s_2=1+3=4$

$s_3=1+3+5=9$

$s_n=1+3+5+\cdots+(2n-1)=n^2$

Fig. A.5

- The *pentagonal* numbers $p_1, p_2, \ldots, p_n$ are the numbers of dots in the regular pentagons (Fig. A.6).
- The *nth tetrahedral number* T_n is the sum of the first n triangular numbers. The dots representing T_n can be arranged in triangular layers, forming a tetrahedron (Fig. A.7).

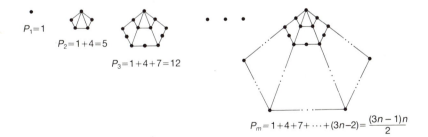

$P_1=1$

$P_2=1+4=5$

$P_3=1+4+7=12$

$P_m=1+4+7+\cdots+(3n-2)=\dfrac{(3n-1)n}{2}$

Fig. A.6

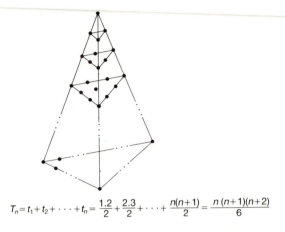

$$T_n = t_1 + t_2 + \cdots + t_n = \frac{1.2}{2} + \frac{2.3}{2} + \cdots + \frac{n(n+1)}{2} = \frac{n\,(n+1)(n+2)}{6}$$

Fig. A.7

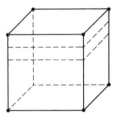

Fig. A.8

- The *nth cube number* C_n is the number of dots in a cube constructed from n layers of n^2 dots, each arranged in a $n \times n$ square (Fig. A.8).

4. **Other familiar numbers and number patterns**

- The *nth Fibonacci number* f_n is the nth member of the so-called Fibonacci sequence: 1, 1, 2, 3, 5, 8, 13, ... In the Fibonacci sequence, starting from $n = 3$, each member is the sum of the previous two:

$$f_n = f_{n-1} + f_{n-2} \qquad \text{for } n = 3, 4, \ldots$$

$$f_2 = f_1 = 1.$$

- A *Pythogorean triple* is a set of three natural numbers a, b, c such that

$$a^2 + b^2 = c^2$$

(e.g. 3, 4, 5 is a Pythagorean triple).
- A *Pythagorean triple* a, b, c is called *primitive* if 1 is the only factor common to a, b and c.
- *Pascal's triangle*, also called *the arithmetic triangle*, is the triangular number pattern shown in Fig. A.9.

$$
\begin{array}{ccccccc}
1 & 1 & 1 & 1 & 1 & 1 & 1 \quad . \quad . \\
1 & 2 & 3 & 4 & 5 & 6 \quad . \quad . \\
1 & 3 & 6 & 10 & 15 \quad . \quad . \\
1 & 4 & 10 & 20 \quad . \quad . \\
1 & 5 & 15 \quad . \quad . \\
1 & 6 \quad . \quad . \\
1 \quad . \quad . \\
. \quad . \\
.
\end{array}
$$

Fig. A.9

Denote the term in the ith (horizontal) row and in the jth (vertical) column of Pascal's triangle by a_{ij}. Then

$$a_{ij} = a_{i+1,j} - a_{i+1,j-1}$$

for all $i = 1, 2, \ldots$ and all $j = 2, 3, \ldots$.
Moreover,

$$a_{ij} = a_{i-1,j} + a_{i-1,j-1} + \cdots + a_{i-1,1}$$

for all $i = 2, 3, \ldots, j = 2, 3, \ldots$.
- The *harmonic triangle* is the triangular number pattern in Fig. A.10. Denote by h_{ij} the term in the ith row and jth column of the harmonic triangle. Then

$$h_{ij} = h_{i-1,j} + h_{i-1,j+1}$$

for all $i = 2, 3, \ldots, j = 1, 2, 3, \ldots$.

$\frac{1}{1}$	$\frac{1}{2}$	$\frac{1}{3}$	$\frac{1}{4}$	$\frac{1}{5}$	$\frac{1}{6}$	.	.
$\frac{1}{2}$	$\frac{1}{6}$	$\frac{1}{12}$	$\frac{1}{20}$	$\frac{1}{30}$	.	.	
$\frac{1}{3}$	$\frac{1}{12}$	$\frac{1}{30}$	$\frac{1}{60}$	.	.		
$\frac{1}{4}$	$\frac{1}{20}$	$\frac{1}{60}$	.	.			
$\frac{1}{5}$	$\frac{1}{30}$	.	.				
$\frac{1}{6}$	.	.					
	.	.					

Fig. A.10

In the harmonic triangle each term h_{ij} is the sum of the infinite series

$$h_{ij} = h_{i+1,j} + h_{i+1,j+1} + h_{i+1,j+2} + \cdots .$$

The sum

$$h_{11} + h_{12} + h_{13} + \cdots = \tfrac{1}{1} + \tfrac{1}{2} + \tfrac{1}{3} + \tfrac{1}{4} + \cdots$$

is the so-called *harmonic series* which diverges. (That is, the sum $\tfrac{1}{1} + \tfrac{1}{2} + \tfrac{1}{3} + \tfrac{1}{4} + \cdots + \tfrac{1}{n}$ does not tend to any finite number as a limit when n tends to infinity.)

• A *matrix M* is a rectangular pattern of symbols:

$$M = \begin{pmatrix} a_{11} & a_{12} & \cdots & a_{1n} \\ a_{21} & a_{22} & \cdots & a_{2n} \\ \vdots & \vdots & & \vdots \\ a_{m1} & a_{m2} & \cdots & a_{mn} \end{pmatrix},$$

denoted briefly as

$$M = (a_{ij})_{\substack{i=1,\ldots,m \\ j=1,\ldots,n}} .$$

We say M is an $m \times n$ matrix.

• Let $A = (a_{ij})_{\substack{i=1,\ldots,m \\ j=1,\ldots,n}}$ and $B = (b_{ij})_{\substack{i=1,\ldots,k \\ j=1,\ldots,\ell}}$

be matrices whose entries are numbers. For such matrices *matrix addition* and *matrix multiplication* are defined in the following special cases according to the rules described below:

- Matrix addition $A + B$ is defined if and only if $m = k$ and $n = \ell$. In that case

$$A + B = (c_{ij})_{\substack{i=1,\ldots,m \\ j=1,\ldots,n}} \qquad \text{where } c_{ij} = a_{ij} + b_{ij}.$$

- Matrix multiplication $A \cdot B$ is defined if and only if $n = k$:

$$A \cdot B = (d_{ij})_{\substack{i=1,\ldots,m \\ j=1,\ldots,\ell}}$$

where

$$d_{ij} = \sum_{t=1}^{n} a_{it} b_{tj} \qquad \text{for all } i = 1, \ldots, m \text{ and all } j = 1, \ldots, \ell.$$

That is, A and B may be multiplied if and only if A is an $m \times n$ and B is an $n \times \ell$ matrix for some m, n, ℓ.

The product of $A = (a_{ij})_{\substack{i=1,\ldots,m \\ j=1,\ldots,n}}$ by a number r is by definition

$$rA = (ra_{ij})_{\substack{i=1,\ldots,m \\ j=1,\ldots,n}}.$$

(b) Polynomials and polynomial equations; diophantine equations

- A *polynomial in one variable x* over the set of real (or complex) numbers is an expression of the form

$$f(x) = a_n x^n + a_{n-1} x^{n-1} + \cdots + a_1 x + a_0,$$

where $a_0, a_1, \ldots, a_n$ are given numbers (real or complex) and n is a given non-negative integer.

$a_0, a_1, \ldots, a_n$ are called the *coefficients* of $f(x)$. If $a_0 = a_1 = \cdots a_n = 0$, then $f(x) = 0$ for all values of x. In this case $f(x)$ is called the *zero polynomial*. For any non-zero polynomial $f(x)$ the *degree* of $f(x)$ is defined as the greatest value of i such that $a_i \neq 0$. If $f(x)$ is of degree $m \geq 1$, then $a_m x^m$ is called the *leading term* of $f(x)$ and a_m is the *leading coefficient* of $f(x)$. A *root* of $f(x)$ is a number α such that $f(\alpha) = 0$.

- A *polynomial in two variables* x, y over the real or complex numbers is an expression of the form

$$f(x, y) = \sum_{r=0}^{n} \sum_{i+j=r} a_{ij} x^i y^j,$$

where a_{ij} are given numbers.

  The numbers a_{ij} are *the coefficients* of $f(x, y)$. If some $a_{ij} \neq 0$ for $i + j = m$ but $a_{ij} = 0$ for all $i + j > m$, then m is called the *degree* of $f(x, y)$.
- Polynomials in three or more variables are defined in a similar way.

(c) Examples of algebraic structures

1. Groups

- An *algebraic operation* on a set S is a rule which assigns to each ordered n-tuple $(a_1, a_2, \ldots, a_n)$ of elements of S a unique element of S. A *binary algebraic operation* on a set S is a rule which assigns to each ordered *pair* (a_1, a_2) of elements of S a unique element of S. For example, addition $(+)$ is a binary algebraic operation on the set $\mathbb{N}$ of natural numbers which associates with each ordered pair (n_1, n_2) of natural numbers their sum $n_1 + n_2$ – a uniquely determined element of $\mathbb{N}$. On the other hand, subtraction is not a binary algebraic operation on $\mathbb{N}$ since the difference of two natural numbers is not necessarily a natural number.
- An *algebraic structure* is a set on which one or more algebraic operations are defined. Groups are among the simplest, and most common algebraic structures.
- A *group* $G(\circ)$ is a non-empty set G on which a binary algebraic operation $\circ$ is defined such that:
  - (i) $a \circ (b \circ c) = (a \circ b) \circ c$ for all a, b, $c \in G$
    (this property of the operation $\circ$ is called *associativity*).
  - (ii) there is an element $e \in G$, called the *identity* of $\circ$, such that

$$a \circ e = e \circ a = a \qquad \text{for all } a \in G;$$

  - (iii) for each element $a \in G$ there is an element $b \in G$ such that

$$a \circ b = b \circ a = e.$$

    b is called the *inverse* of a with respect to $\circ$.

- A *commutative group* is a group $G(\circ)$ such that

$$a \circ b = b \circ a \qquad \text{for all } a, b \in G.$$

Commutative groups are also called *abelian groups*.

Examples of commutative groups: (i) The set $\mathbb{Z}$ of integers with respect to addition. (ii) The set $\{1, 2, 3, 4\}$ with respect to multiplication modulo 5. (The product of two integers modulo 5 is the remainder of the usual product of the integers when divided by 5, e.g. the product of 3 and 4 modulo 5 is 2, since $3 \times 4 = 12$ and the remainder of 12 when divided by 5 is 2.)

Examples of non-commutative groups: (i) The set of 2×2 matrices

$$\begin{pmatrix} a & b \\ c & d \end{pmatrix},$$

where a, b, c, d are real numbers such that $ad - bc \neq 0$, with respect to matrix multiplication. (ii) The set of quaternions different from 0, with respect to multiplication of quaternions (see Chapter III, Section 3).
- A *finite group* is a group with finitely many elements.
- The *order of a finite group* is the number of elements in the group.
- The *order of an element* g is a group $G(\circ)$ is the smallest non-negative integer k such that

$$\underbrace{g \circ g \circ \cdots \circ g}_{k \text{ times}} = e,$$

where e is the identity element of G. If such an integer k does not exist, that is if

$$\underbrace{g \circ g \circ \cdots \circ g}_{n \text{ times}} \neq e$$

for any non-negative integer n, then g is said to be of *infinite order*.
- An *elementary abelian group* is an abelian group $G(\circ)$ with the following property: There is a prime number p such that every non-identity element of G is of order p. It is easy to verify that in that case the order of G is a power of p.

  Problem 111 refers to an elementary abelian group of order 16. In this group the non-identity elements are of order 2.
- A set of elements $\{g_1, g_2, \ldots, g_k\}$ of a finite abelian group $G(\circ)$ is called a

set of *generators* of G if every element of G can be written in the form $g_1^{n_1}$ $\circ\ g_2^{n_2} \circ \cdots \circ g_k^{n_k}$, where $n_i \in \{0, 1, 2, \ldots\}$,

$$g_i^{n_i} = \{\underbrace{g_i \circ g_i \circ \cdots \circ g_i}_{n_i \text{ times}}\},$$

and $g_i^0 = e$.

2. Real vector spaces

• A *real vector space* V is a commutative group $V(\oplus)$ with respect to an operation $\oplus$ (called vector addition) such that between the real numbers and the elements of V an external operation $\odot$ is defined, satisfying the rules:

(i) $r \odot v = V$ for all real numbers r and all $v \in V$;
(ii) $(r_1 + r_2) \odot v = r_1 \odot v + r_2 \odot v$ for all real numbers r_1, r_2 and all $v \in V$;
(iii) $r \odot (v_1 + v_2) = r \odot v_1 + r \odot v_2$ for all real numbers r and all v_1, $v_2 \in V$.
(iv) $r_1 \odot (r_2 \odot v) = (r_1 r_2) \odot v$ for all real numbers r_1, r_2 and all $v \in V$, and
(v) $1 \odot v = v$ for all $v \in V$.

The elements of V are called *vectors* and the real numbers are often referred to as *scalars*.
Examples of real vector spaces: (i) The set $\mathbb{R}^2$ of ordered pairs (x_1, x_2) of real numbers x_1, x_2. In $\mathbb{R}^2$ vector addition is defined by

$$(x_1, x_2) \oplus (x_1', x_2') = (x_1 + x_1', x_2 + x_2') \qquad \text{for all } x_i, x_i' \in \mathbb{R}, i = 1, 2.$$

and the external operation between vectors and scalars is defined by

$$r \odot (x_1, x_2) = (rx_1, rx_2) \qquad \text{for all } r, x_1, x_2 \in \mathbb{R}.$$

The vector (x_1, x_2) has a geometric representation in the two-dimensional Cartesian coordinate system shown below (for the definition of Cartesian coordinate systems see (d)).

$(x_1, x_2) \in \mathbb{R}^2$ is represented by the oriented line segment **OP** starting at the origin O of the coordinate system and ending at the point P with Cartesian coordinates $x = x_1$ and $y = x_2$. The distance $OP = \sqrt{(x_1^2 + x_2^2)}$ is called *the length* of the vector **OP**. Moreover, with any oriented line segment of the plane xOy, starting at a point A and ending at a point B, there is associated a vector **AB** (Fig. A.11). The vector **AB** is defined to

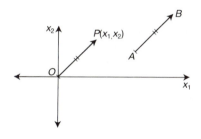

Fig. A.11

be equal to the vector **OP** if and only if the oriented line segments *OP* and *AB* are on parallel lines, point in the same direction, and have equal lengths.

(ii) The generalization of $\mathbb{R}^2$ is $\mathbb{R}^n$ for any natural number n. The real vector space $\mathbb{R}^n$ is the set of ordered n-tuples $(x_1, x_2, \ldots, x_n)$, where $x_i \in \mathbb{R}$ for $i = 1, 2, \ldots, n$. Vector addition and multiplication by scalars are defined according to the rules:

$$(x_1, x_2, \ldots, x_n) \oplus (x'_1 + x'_2 + \cdots + x'_n) = (x_1 + x'_1, x_2 + x'_2, \ldots, x_n + x'_n)$$

and

$$r \odot (x_1, x_2, \ldots, x_n) = (rx_1, rx_2, \ldots, rx_n)$$

for all $x_i, x'_i, r \in \mathbb{R}, i = 1, 2, \ldots, n$.

The vectors (x_1, x_2, x_3) of $\mathbb{R}^3$ are represented in the three-dimensional Cartesian coordinate system as follows.

The vector (x_1, x_2, x_3) is represented by the oriented line segment **OP**, where O is the origin of the coordinate system and P is the point with coordinates $x = x_1$, $y = x_2$ and $z = x_3$. The length of **OP** is

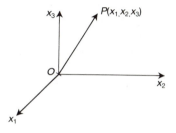

Fig. A.12

$\sqrt{(x_1^2 + x_2^2 + x_3^2)}$. For any two points A, B in this coordinate system the oriented line segment with start A and endpoint B represents a vector. $AB = \mathbf{OP}$ if and only if the oriented segments AB and OP are on parallel lines, point in the same direction and have equal lengths. The representation of the vectors $(x_1, x_2, \ldots, x_n) \in \mathbb{R}^n$ for any natural number n is done in a similar way in an n-dimensional Cartesian coordinate system.

In the last part of Section (c) our aim is to explain the dimension of a vector space.

- A *linear combination* of k vectors $\mathbf{v}_1, \mathbf{v}_2, \ldots, \mathbf{v}_k$ of a real vector space $V(\oplus)$ is an expression of the form
  $r_1 \odot \mathbf{v}_1 \oplus r_2 \odot \mathbf{v}_2 \oplus r_3 \odot \mathbf{v}_3 \oplus \cdots \oplus r_k \odot \mathbf{v}_k$, where r_i are real numbers, $i = 1, 2, \ldots, k$.
- A set $S = \{\mathbf{v}_1, \mathbf{v}_2, \ldots, \mathbf{v}_k\}$ of vectors from V is said to be *linearly independent* if the only linear combination $r_1 \odot \mathbf{v}_1 \oplus r_2 \odot \mathbf{v}_2 \oplus \cdots \oplus r_k \odot \mathbf{v}_k$ equal to the identity element of $V(\oplus)$ (usually denoted by $\mathbf{O}$) is the one in which $r_i = 0$ for all $i = 1, 2, \ldots, k$. Otherwise the set S is called *linearly dependent*.
- A *basis* of a real vector space V is a set $\mathbb{B} \subseteq V$ of linearly independent vectors such that any vector of V is a linear combination of finitely many elements from $\mathbb{B}$.
  Example of a basis: The vectors $(1,0)$ and $(0,1)$ represent a basis of $\mathbb{R}^2$, since:

$$r_1(1,0) + r_2(0,1) = (0,0)$$

implies that $r_1 = r_2 = 0$, and

$$(x_1, x_2) = x_1(1,0) + x_2(0,1)$$

for any $x_1, x_2 \in \mathbb{R}$.

A vector space can have many bases. For example, $\mathbb{B}' = \{(1,1), (-1,1)\}$ is also a basis for $\mathbb{R}^2$. (Verify it!) It can be proved (see [61]) that if one basis $\mathbb{B}$ of a vector space V over F has finitely many elements — say, m — then any other basis of V has the same number m of elements. The number m is called the *dimension* of V. There are vector spaces whose bases consist of infinitely many vectors; we shall not consider such vector spaces.

The vector space $\mathbb{R}^n$, defined above, is n-dimensional.

(d) Coordinate systems

1. Cartesian coordinate systems

- A *one-dimensional Cartesian coordinate system* is a real number line ℓ (Fig. A.13).

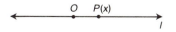

Fig. A.13

   With any point P on ℓ there is associated *one* coordinate x; x is the real number representing P on the number line ℓ (see p. 207).
- A *two-dimensional Cartesian coordinate system* consists of two orthogonal real number lines meeting at their common origin O (Fig. A.14). The point O is called the origin of the system and the number lines are called the axes of the system. Label the axes x and y. Any point P in the plane carrying the coordinate axes has two coordinates x, y defined as follows. Drop perpendiculars from P onto the x and the y-axis. Denote the feet of these perpendiculars by P' and P'' respectively. The points P' and P'' have their coordinates in the one-dimensional coordinate systems on the number lines of the axes. Let the coordinate of P' in the one-dimensional coordinate system on the x-axis be x, and the coordinate of P'' on the one-dimensional coordinate system on the y-axis by y. Then the two coordinates x and y are associated (x as the first, y as the second coordinate) with P in the two-dimensional coordinate system. In the two-dimensional Cartesian coordinate system, the coordinates of the points P' and P'' are x, 0 and 0, y respectively.

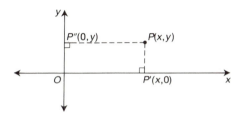

Fig. A.14

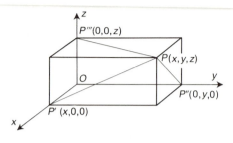

Fig. A.15

- A *three-dimensional Cartesian coordinate system* consists of three mutually orthogonal real number lines meeting at their common origin. Let us call the axes x, y and z. An arbitrary point P in space has three coordinates x, y, z defined as follows:

  Drop perpendiculars from P onto the x, y and z axes. Denote the feet of the perpendiculars by P', P'' and P''' respectively. In the one-dimensional coordinate systems on the x, y and z axes P', P'', P''' have their coordinates x, y and z respectively.

  In the three-dimensional coordinate system, P has the coordinates x, y, z. (Thus in the space the points P', P'', P''' have the coordinates $x,0,0$, $0,y,0$ and $0,0,z$ respectively.)

2. Barycentric coordinate systems

in the plane and in space are described in the solution of Problem 55.

3. Argand diagrams

Argand diagrams are used for geometric representation of complex numbers. An Argand diagram consists of two orthogonal axes: the real x-axis carrying the real number line, and the imaginary y-axis carrying the imaginary number line (that is, a number line in which the numbers associated with points are of the form iy, where $y \in \mathbb{R}$). The two axes meet at their common origin (Fig. A.16).

A complex number $z = x + iy$ is represented on the Argand diagram by a point P such that:

- the foot P' of the perpendicular from P onto the real axis has coordinate x on the real number line;

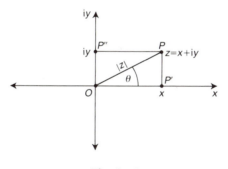

Fig. A.16

- the foot P'' of the perpendicular from P onto the imaginary axis has coordinate iy on the imaginary number line.

The distance $OP = \sqrt{(x^2 + y^2)}$ is called the *modulus* $|z|$ of the real number z; the angle, measured counterclockwise, between the positive direction of the x-axis and the straight line segment OP is called the *argument* θ of z.

(e) Convex point sets; polygons; inscribed and circumscribed circles

- A set S of points (in the plane, or in the space) is called *convex* if with any two points A, B in S all points of the straight line segment AB are in S.
- Let us define a *polygon in the plane* as a point set bounded by a closed polygonal line $A_1A_2 \cdots A_nA_1$. The *closed polygonal line* $A_1A_2 \cdots A_nA_1$ consists of the sequence of straight line segments A_1A_2, A_2A_3, ..., $A_{n-1}A_n$, A_nA_1 connecting the points $A_1, A_2, \ldots, A_n$. The points $A_1, A_2, \ldots, A_n$ are the vertices and the straight line segments $A_1A_2, \ldots, A_nA_1$ are the sides of the polygon.
- A *polygon* is called *convex* if the set of points inside and on the boundary of the polygon is convex.
- A *non*-convex polygon is shown in Fig. A.17.
- A *circle c* is said to be *insribed* in a polygon P if all sides of P are tangent to c. The sides of P are the straight line segments joining the pairs of consecutive vertices of P. Thus circle c in Fig. 18(a) is inscribed in $\triangle ABC$, while circle c' in Fig. A.18(b) is not inscribed in $\triangle DEF$.
- A *circle c* is *circumscribed* about a polygon if all vertices of the polygon are on c (Fig. A.19).

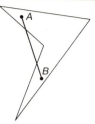

Fig. A.17

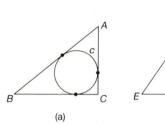 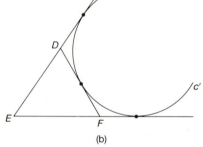

(a)   (b)

Fig. A.18

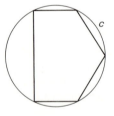

Fig. A.19

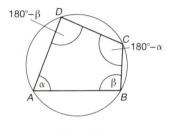

Fig. A.20

- A *cyclic quadrilateral* is a quadrilateral whose vertices lie on a common circle (Fig. A.20). In a cyclic quadrilateral opposite angles add up to 180° ($\angle ABC + \angle ADC = 180°$ and $\angle DAB + \angle DCB = 180°$. (Prove it!)

(f) Geometric transformations (rotation, inversion, stereographic projection, translation, reflection)

There are different types of geometric transformations. Here we are concerned with mappings of point sets onto point sets in a plane, or in space. Different geometric transformations may preserve different properties of the shapes (i.e. angles, parallelism or lengths of sides). Sometimes it is useful to apply a particular transformation to obtain a simpler situation in which a particular property of a problem is preserved.

1. Rotation about a point

A rotation about a point is a mapping of the point set $\mathbb{P}$ of a plane π onto itself defined as follows.

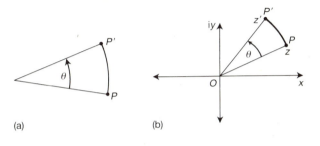

(a)   (b)

Fig. A.21(a)   **Fig. A.21(b)**

Let O be a fixed point of $\mathbb{P}$ and let θ be a given angle. In π the *rotation about O through θ* is the mapping μ associating with each point $P \in \mathbb{P}$ the point P' such that

$$OP = OP' \quad \text{and} \quad \measuredangle POP' = \theta$$

where θ is measured counterclockwise (Fig. A.21(a)). Using complex numbers we can describe μ algebraically as follows:

An Argand diagram is set up in π with origin O (Fig. A.21(b)). Let z be the complex number assigned to P in this diagram. Then the complex number z' assigned to the image P' of P under μ is the product $z(\cos \theta + i \sin \theta)$. (Verify this!)

2. Rotation about an oriented axis

A rotation about an oriented axis is a mapping of the point set S in three-dimensional space onto itself, defined as follows.

Let ℓ be an oriented straight line in space and let θ be a given angle. The rotation about ℓ through θ is the mapping ν such that the image P' of an arbitrary point $P \in S$ has the properties:

- $PR = P'R$, where R is the point of intersection of ℓ with the plane through P perpendicular to ℓ;
  $\measuredangle PRP' = \theta$;
- the oriented line segments RP, RP' and the oriented axis ℓ conform to the 'right-hand rule' (Fig. A.22(a)).

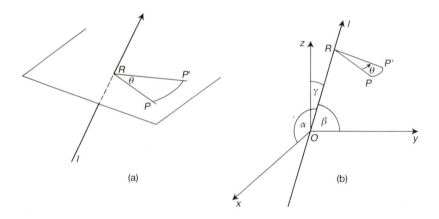

(a)   (b)

Fig. A.22

v can be described algebraically using quaternions, as follows.

Set up a three-dimensional Cartesian coordinate system in space with origin O and ℓ, such that the positive direction of the z-axis makes an angle $\gamma \leq 90°$ with the directed line ℓ. Denote the angles of the directed line ℓ with the positive direction of the x- and y-axes by α and β respectively. To an arbitrary point $P(x, y, z)$ in space assign the quaternion $xi + yj + zk$. Then the quaternion $x'i + y'j + z'k$ assigned to the image $P'(x', y', z')$ of P under v is given by

$$x'i + y'j + z'k = q(xi + yj + zk)\bar{q},$$

where

$$q = \cos\frac{\theta}{2} + (i\cos\alpha + j\cos\beta + k\cos\gamma)\sin\frac{\theta}{2}$$

and

$$\bar{q} = \cos\frac{\theta}{2} - (i\cos\alpha + j\cos\beta + k\cos\gamma)\sin\frac{\theta}{2}.$$

The proof is left to the reader.

3. Inversion with respect to a circle

An inversion with respect to a circle c with centre O and radius r is a transformation σ, mapping an arbitrary point $P \neq 0$ in the plane of c onto the point P' such that:

- P' is on the straight line OP, on the same side of O as P; and the product of the distances OP and OP' is $OP \cdot OP' = r^2$.

  From the above definition it is clear that if P' is the image of P under σ, then the image of P' under σ is P. Figure A.23 shows a simple method of constructing images of points under σ.

If A is a point outside c, draw a tangent from A to c, and from the point of tangency T drop the perpendicular onto OA. The foot A' of this perpendicular is the image of A under σ. If B is a point inside c, draw a perpendicular to OB through B, intersecting c at T'. At T' draw the tangent to c and mark its point of intersection with the straight line OB with B'. The point B' is the image of B. If a point P is on c then $P' = P$. The centre O of c has no image under σ.

It is a simple exercise to prove that under σ the image

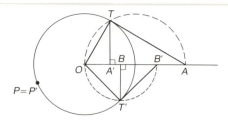

Fig. A.23

(i)   of a straight line ℓ through O is the same line ℓ;
(ii) of a circle c through O is a straight line not through O;
(iii) of a straight line not through O is a circle through O;
(iv) of a circle not through O is a circle not through O.

4. Inversion with respect to a sphere

An inversion with respect to a sphere S with centre O and radius r is a transformation κ mapping an arbitrary point $P \neq 0$ in space onto the point P' such that:

- P' is on the straight line OP, on the same side of O as P; and
- the product of the distances OP and OP' is $OP \cdot OP' = r^2$.

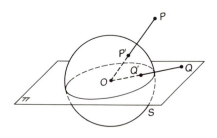

Fig. A.24

Let π be an arbitrary plane through O. π intersects S along a so-called great circle of S, say c, with centre O and radius r. Call σ the inversion of π respect to c. The definition of κ implies that the image Q' of any point $Q \neq 0$ in π under κ is the same as the image of Q under σ. In other words:

If π is any plane through O, then the inversion κ with respect to the sphere S induces in π an inversion with respect to the circle along which π intersects S.

Using the above statement one can prove that under κ the image:

(i)   of a plane K through O is the same plane K;
(ii) of a sphere Σ through O is a plane not through O;
(iii) of a plane K not through O is a sphere through O;
(iv) of a sphere Σ not through O is a sphere not through O.

(*Hint*: In cases (i) and (iii) draw the perpendicular p from O to K, passing through O, and in cases (ii) and (iv) draw the straight line p joining O to the centre of Σ. For any plane π through p denote its intersection with S by c_π, and the intersection of π with K or Σ by K_π or Σ_π respectively. Since the point sets of K, or Σ, are unions of the point sets of K_π or Σ_π respectively, when π revolves around p, it follows that the images of π and Σ are the unions of the images of K_π or Σ_π respectively under the inversions with respect to the corresponding circles c_π.)

5. Stereographic projection

A stereographic projection is a mapping λ of the set of points of a sphere S except for a point C on S — the centre of the projection — onto the set of points of a plane π not passing through C. We shall consider the case when π is tangent to S at the point C' diametrically opposite to C. The image P' of any point $P \neq C$ under λ is the intersection of the straight line CP with π. λ has the following important properties:

(i) λ maps circles on S passing through C onto straight lines;
(ii) λ maps circles on S which do not pass through C onto circles.

(i) and (ii) can be proved by applying inversion with respect to the sphere S' with centre C and radius CC'. Call this inversion κ. Triangles CPC' and $CC'P'$ are similar; hence $CP \cdot CP' = (CC')^2$ (see Fig. A.25). It follows that P' is the image of P under κ. For any circle c on S denote the plane carrying c by Σ. The inversion κ maps S (except C) onto π and it maps Σ onto itself if Σ

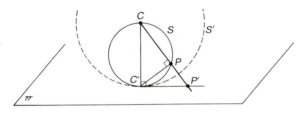

Fig. A.25

passes through C. In that case the image c' of c under κ is the intersection of Σ with π, that is a straight line. If Σ does not pass through C then it is mapped under κ onto a sphere Σ'. Hence c' is the intersection of the sphere Σ' with the plane π, which is a circle.

6. Translation

A translation with translation vector **v** is a mapping λ of the point set $\mathbb{P}$ of a plane or of space onto itself such that the image of an arbitrary point $A \in \mathbb{P}$ is the end-point A' of the vector **AA′** = **v** (Fig. A.26).

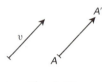

Fig. A.26

7. Reflection

A reflection in a straight line ℓ in a plane π is a mapping α of the set of points $\mathbb{P}$ of π onto itself defined as follows.

- α maps an arbitrary point $P \in \mathbb{P}$ onto P' such that
- PP' is perpendicular to ℓ and
- the intersection of the straight line PP' with ℓ is the midpoint M of the line segment PP' (Fig. A.27(a)).

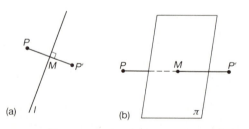

Fig. A.27(a)   **Fig. A.27(b)**

A reflection in a plane π is a mapping of the set of points of space onto itself such that:

- the image P' of an arbitrary point $P \in \mathbb{P}$ is on the straight line through P perpendicular to π; and
- the perpendicular PP' intersects π in the midpoint M of the line segment PP' (Fig. A.27(b)).

(g) Combinatorial concepts

1. Permutations, combinations, binomial coefficients

- *A permutation* of n different objects is an arrangement of the elements into a sequence. The order of the elements in the sequence matters. (Thus the permutation 132 of the objects 1, 2, 3 differs from the permutation 123,) The number of all permutations of n different objects is $n(n-1)(n-2)\cdots3\cdot2\cdot1$. This number is denoted by $n!$ By definition $0! = 1$.

  A permutation of r different objects chosen from a set S of n different objects is an ordered arrangement of r objects chosen from the set S. The number of permutations of r different objects chosen from S is $n(n-1)(n-2)\cdots[n-(r-1)]$, that is $n!/(n-r)!$. Denote this number by P_r^n.

- A *permutation with repetition* is a permutation in which objects selected need not be different (that is, the same object can appear repeatedly). The number of permutations with repetitions of r different objects chosen from n objects is

$$\underbrace{n \cdot n \cdots \cdots n}_{r \text{ times}} = n^r;$$

  this is so because there are n choices for each of the r elements (e.g. there are $3^2 = 3 \times 3$ permutations with repetitions of 2 different objects chosen from the set $\{a, b, c\}$; these are $a, a; a, b; a, c; b, a; b, b; b, c; c, a; c, b; c, c$.

- A *combination* is an unordered selection of objects from a set. The number C_r^n of combinations of n objects taken r at a time, without repetition, is equal to $P_r^n/r!$. Thus $C_r^n = n!/(n-r)!r!$. The number C_r^n is often denoted by $\binom{n}{r}$.

- *The number of permutations of n objects, of which r_1 are of one kind, r_2 of a second kind, . . ., r_k of a kth kind*, where obviously

$$r_1 + r_2 + \cdots + r_k = n,$$

is

$$C^n_{r_1, r_2, \ldots, r_k} = \frac{n!}{r_1! r_2! \ldots r_k!}.$$

(This result is obtained by noticing that if the r_i objects of ith kind are provisionally considered to be different for all i, then there are $n!$ permutations of the n objects. But by identifying the like objects each permutation is counted $r_1! r_2! \ldots r_k!$ times.)

- *The binomial coefficients* are the numbers $\binom{n}{r}$ for any pair of non-negative integers n, r such that $n \geq r$. They are called this because they appear as coefficients in the expansion of the binomial $(x+y)^n$ for $n = 0, 1, 2, \ldots$. The binomial coefficients satisfy the recursive formula

$$\binom{n+1}{k} = \binom{n}{k-1} + \binom{n}{k} \text{ for } n \geqslant k \geqslant 1;$$

$$\binom{n}{0} = \frac{n!}{0! n!} = 1 \text{ for } n \geqslant 1,$$

and

$$\binom{0}{0} = \frac{0!}{0!} = 1.$$

Hence they are the entries of Pascal's triangle.
- *The trinomial coefficients* are the numbers $n!/r_1! r_2! r_3!$ where r_1, r_2, r_3 are non-negative integers adding up to an integer $n \geq 0$. They are the coefficients in the expansion of the trinomial $(x + y + z)^n$.
- *The multinomial coefficients* are generalizations of the binomial and trinomial coefficients. These are the numbers $n!/r_1! r_2! \cdots r_k!$ where n is a non-negative integer, k is a natural number, and $r_1, r_2, \ldots, r_k$ are non-negative integers adding up to n. The multinomial coefficients are the coefficients in the expansion of the polynomial $(x_1 + x_2 + \cdots + x_k)^n$.

2. Graphs

A *graph G* is a set of points, called *vertices*, and a set of *edges*. Each edge joins two vertices, called the *endpoints* of the edge. There may be no edge joining two vertices, or one edge, or more. An edge is called a *loop* if its endpoints coincide.

A *path* of G is a succession of edges $a_1, a_2, \ldots, a_k, a_{k+1}, \ldots$ connecting a succession of vertices $P_1, P_2, \ldots, P_k, \ldots$; that is, a_1 connects P_1 and P_2, a_2 connects P_2 and $P_3, \ldots, a_k$ connects P_k and $P_{k+1}, \ldots$. We say that the path passes through $P_1, P_2, \ldots$. Two vertices of a graph are said to be *connected* if there is path passing through them.

A *connected graph* is a graph in which every pair of vertices is connected.

An edge of a graph is called *oriented* if it is given a direction pointing from one of its endpoints, say A, to the other, say B. The vertex A is then called the beginning, and B the end of the edge.

An *oriented* graph is a graph in which every edge is oriented.

For a path consisting of the edges $a_1, a_2, \ldots, a_k, a_{k+1}$ *in an oriented graph*, it is required that the endpoint of a_i is the beginning of a_{i+1} for all $i = 1, 2, \ldots$.

A path in an oriented graph consisting of the edges $a_1, a_2, \ldots, a_n$ is called a *cycle* if the endpoint of a_n is the beginning of a_1.

3. Balanced incomplete block designs

The block designs investigated in Chapter IV, Section 1.4 are weaker versions of combinatorial structures known in the literature as balanced incomplete block designs. (The term 'balanced incomplete block design' comes from the theory of design of experiments, a branch of statistics.)

A *balanced incomplete block design* $\mathscr{D}$ is an arrangement of a set of v distinct objects, called *points*, into subsets called *blocks* satisfying the following conditions:

- each block contains k points
- each point is contained in r blocks
- every pair of distinct points occurs together in λ blocks.

The number b of blocks in $\mathscr{D}$, and the numbers v, k, r and λ are called the parameters of $\mathscr{D}$.

Examples of a balanced incomplete block design.

Projective planes with finitely many points (defined in Chapter III, Section 4.2). The parameters of these designs are:

$$k = r = n + 1, \qquad v = b = n^2 + n + 1, \qquad \lambda = 1$$

where n is a natural number greater than 1.

(Conversely, it can be proved that a balanced incomplete block design with parameters $k = r = n + 1$, $v = b = n^2 + n + 1$, $\lambda = 1$, where n is a natural number greater than 1, is a projective plane.)

In the above examples $v = b$. In general $b \geq v$. This result, proved by Fisher for balanced incomplete block designs, was extended by Hanani to a wider class of structures, which in this book (Chapter IV, Section 1.4) are called simply *block designs*. In these structures each pair of distinct points occurs in one block, every block has at least two points and $b > 1$, but the number of points per block, and the number of blocks per point need not be the same for all blocks, respectively points.

(h) Number sequences and limits

• Let $f(x)$ be a function defined for every natural number n. The ordered set of values

$$f(1), f(2), f(3), f(4), \ldots, f(n), \ldots$$

is called an *infinite number sequence*. Its elements are often denoted by $a_1, a_2, a_3, a_4, \ldots, a_n, \ldots$ respectively.

Examples of infinite number sequences:
(i)   The function $f(n) = n/(1 + n)$ defines the number sequence

$$\tfrac{1}{2}, \tfrac{2}{3}, \tfrac{3}{4}, \tfrac{4}{5}, \ldots$$

(ii) The function $f(n) = (-1)^n\, 1/n$ defines the number sequence

$$-1, \tfrac{1}{2}, -\tfrac{1}{3}, \tfrac{1}{4}, \ldots$$

(iii)   The function $f(n) = n^2$ defines the number sequence

$$1, 4, 9, 16, 25, 36, \ldots$$

• We say that *a number sequence $a_1, a_2, \ldots$ tends to infinity* (symbolically $a_n \to \infty$) if for any positive number c there exists a natural number N such that $a_{N+1}, a_{N+2}, \ldots$ will be greater than c.
• *A number sequence $a_1, a_2, a_3, \ldots, a_n$ has the limit ℓ as n tends to infinity* if and only if for any positive number ϵ, no matter how small, there exists a positive integer N (depending on ϵ), such that

$$|\ell - a_n| < \epsilon \qquad \text{for all integers } n \geq N.$$

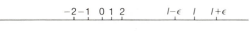

Fig. A.28

Here $|\ell - a_n|$ denotes the absolute value of the difference $\ell - a_n$. (The absolute value $|x|$ of a real number x is equal to x if $x \geq 0$ and to $-x$ if $x < 0$.)

In other words, the terms $a_N, a_{N+1}, a_{N+2}, \ldots$ of the number sequence $a_1, a_2, \ldots$ are all in the interval between the numbers $\ell - \epsilon$ and $\ell + \epsilon$ on the number line (Fig. A.28).

Examples:

(i) $\displaystyle\lim_{n \to \infty} \frac{n}{1 + n} = 1.$

(ii) $\displaystyle\lim_{n \to \infty} (-1)^n \frac{1}{n} = 0.$

(iii)   The sequence $1^2, 2^2, 3^2, 4^2, \ldots$ does not converge to a finite limit as n tends to infinity. Instead $n^2 \to \infty$ as $n \to \infty$.

Appendix II Notes on mathematicians mentioned in the text

ABEL, Niels Henrik (1802–29), Norwegian mathematician. Despite his early death, Abel achieved profound results in Algebra and in the Theory of Functions. He gave the first satisfactory proof of the unsolvability 'in radicals' of the general quintic equation

$$ax^5 + bx^4 + cx^3 + dx^2 + ex + f = 0.$$

(In other words, he showed that there are no formulae expressing the solutions of the equation.)

ARCHIMEDES of Syracuse (circa 287–212 B.C.) is considered to be the founder of mathematical physics: from a set of simple facts he deduced conclusions establishing significant relationships between mathematics and mechanics. Apart from his famous work on the equilibrium of planes and properties of floating bodies, Archimedes composed a series of remarkable mathematical treatises: on spirals, on the sphere and cylinder, on the quadrature of parabolic segments and on the measurement of a circle. He was also interested in elementary problems; his *Book of Lemmas* contains a study of the shoemaker's knife (Problem 89).

BOLYAI, János (1802–60), a Hungarian mathematician. Independently of Lobachevsky, Bolyai constructed the geometry, nowadays known as the Geometry of Bolyai–Lobachevsky, or hyperbolic geometry. This was the first geometry in the history of mathematics in which the Euclidean parallel postulate did not hold.

DE BRUIJN, N.G. Contemporary Dutch mathematician.

BUFFON, Comte de, Georges Louis Leclerc (1707–88), French naturalist and writer, was the famous author of the multivolume *Histoire Naturelle*, and the founder of the botanical gardens in Paris. The theory of probability is indebted to him for pointing out that problems in probability can often be expressed in geometric form and solved by geometric methods.

CAYLEY, Arthur (1821–95), celebrated English mathematician, was rivalled in productivity perhaps only by Euler and Cauchy. He took an active part in the creation of modern algebra, and the theory of algebraic structures. Cayley started the study of matrices in connection with his work on the transformation of coordinates.

COPERNICUS, Nicholas (1473–1543), Polish astronomer, revolutionized the world view by 'putting the earth in motion about the sun'. He was versed in trigonometry and made a number of minor contributions to mathematics.

DESARGUES, Girard (1591–1661) was a French architect and military engineer. He may well be said to have been the founder of projective geometry. Due to his unusual terminology and the triumph of Descartes' analytic method, Desargues' contribution was forgotten. It was rediscovered in the nineteenth century when projective geometry became a flourishing discipline. The theorem of Desargues plays a key role in projective geometry.

DESCARTES, René (1596–1650), French philosopher and mathematician, was the founder of analytic geometry. The aim of this discipline is twofold: one is the application of algebra to geometry and the other is the translation of algebraic operations into geometric language. The key element of this approach was Descartes' coordinatization of the plane, subsequently extended to n-dimensional spaces for any natural number n.

The polyhedral formula $v + f = e + 2$, where v, f, and e denote the number of vertices, faces and edges of a polyhedron, usually named for Euler, was in fact discovered by Descartes.

DIOPHANTUS of Alexandria lived in the third century A.D. His chief treatise, *Arithmetica*, a collection of algebraic problems, contains examples of indeterminate equations. Skilful solutions, described in the treatise, influenced the emergence and development of the theory of Diophantine equations, that is of equations with integer coefficients whose solutions are rational numbers (or integers). Diophantus is regarded as the first European algebraist.

DIRICHLET, Peter Gustav Lejeune (1805–59), German mathematician, initiated research in analytic number theory, that is the application of analysis to number-theoretic problems.

ERDÖS, Paul, a celebrated contemporary mathematician of Hungarian origin, is well known for formulating and solving problems from a wide range of topics from the theory of numbers, set theory, combinatorics, theory of designs, graph theory, theory of probability, and analysis. Himself a child prodigy, Erdös takes a serious interest in advancing mathematically gifted youngsters.

EUCLID of Alexandria (around 300 B.C.) was the author of the most successful mathematics textbook ever written: the *Elements*. This is an introduction to elementary mathematics: geometry, algebra and number theory. Euclid's exposition relies on the axiomatic method (see Chapter III, Section 4.1), the greatest Greek contribution to mathematics as well as to all organized thought.

EULER, Leonhard (1707–83), born in Basel, was one of the most significant mathematicians of all times. He added knowledge to virtually every branch of pure and applied mathematics of his time. Euler's interests ranged from the most elementary to the most advanced problems. Apart from important mathematical results, posterity is indebted to Euler for taking an active part in the shaping of mathematical language and in modernizing notation. Some of the symbols introduced by Euler are e, i and π.

FALTINGS, Gerd, contemporary German mathematician. Faltings' solution of the Mordell conjecture is one of the great achievements of twentieth-century mathematics.

FERMAT, Pierre de (1601–65), a French lawyer and councillor at the local *parlement* in Toulouse, was the greatest mathematical amateur in history. Apart from famous work on number theory, Fermat made fundamental contributions to analytic geometry and analysis.

FIBONACCI, son of Bonaccio (1170–1250), was the nickname of Leonardo of Pisa. His celebrated book *Liber abaci* is a treatise on algebraic methods and problems. It recommends the use of Hindu–Arabic numerals instead of Roman numerals. The problem in *Liber abaci* that has most inspired future mathematicians was: How many pairs of rabbits will be produced in a year, beginning with a single pair, if in every month each pair bears a new pair which becomes productive from the second month on? This celebrated problem gave rise to the Fibonacci sequence, 1, 1, 2, 3, 5, 8, 13, 21, ... The Fibonacci sequence has many fascinating properties and appears in connection with various topics in mathematics and science.

GALLAI, Tibor, twentieth-century Hungarian mathematician.

GAUSS, Carl Friedrich (1777–1855) was an infant prodigy. As a ten-year-old boy he discovered a quick way of calculating the sum

$$S_{100} = 1 + 2 + \cdots + 100$$

by proving that $S_{100} = \frac{1}{2} \times 100 \times 101$. At the age of fifteen, starting his university studies in Göttingen, Gauss was unsure whether to become a philologist or a mathematician. However, on March 30, 1796, he made a brilliant mathematical discovery, prompting him to devote his life to mathematics. On that day Gauss carried out the construction of a regular polygon with 17 sides by using ruler and compasses.

Gauss described mathematics as the queen of sciences and the theory of numbers as the queen of mathematics. Among his profound results in number theory are the fundamental theorem of arithmetic and the theory of congruences (see [60]). Gauss' doctoral thesis contained the proof of the fundamental theorem of algebra (see p. 140). He later extended his result on the construction of a regular 17-gon by stating which regular polygons were constructible with ruler and compasses. Gauss also made important contributions to analysis, astronomy, statistics and geometry.

GREGORY, James (1638–75), Scottish mathematician, is regarded as a predecessor of Newton because of his important work on infinite processes. The series of arc tan x (Chapter III, Section 2) bears his name.

HAMILTON, William Rowen (1805–65) was a celebrated Irish mathematician and scientist. His interests in applying mathematics to physics led to the creation of quaternions (Chapter III, Section 3).

HANANI, Haim, contemporary Israeli mathematician.

HERON of Alexandria (about 100 A.D.) is best known for the formula bearing his name, for calculating the area A of a triangle in terms of its side lengths a, b, c: $A = \sqrt{[s(s-a)(s-b)(s-c)]}$, where $S = \frac{1}{2}(a + b + c)$. Heron was interested in mensuration in various disciplines: geometry, optics, mechanics, and geodesy. By applying a simple geometric argument (see Problem 69) Heron proved that the path of a light ray travelling from a source to a mirror and then to the observer is the shortest possible. Heron is remembered in the history of science as the inventor of a primitive steam engine, of a forerunner of the thermometer and of various mechanical toys.

HILBERT, David (1862–1943), versatile German mathematician, contributed to the theory of numbers, mathematical logic, differential equations and to mathematical physics. In his celebrated book *Grundlagen der Geometrie (Foundations of Geometry)*, Hilbert sharpened the axiomatic method of Euclid and corrected its shortcomings. Through this work Hilbert became the leader of an 'axiomatic school', which exerted a strong influence on modern mathematics.

At the Second International Mathematical Congress held in 1900 in Paris, Hilbert gave a lecture in which he listed 23 problems which he believed would be, or should be, among those occupying the attention of research workers in the twentieth century. These problems, many of which are still unsolved, became famous as 'Hilbert's 23 problems'.

HURWITZ, Adolf (1859–1919), German mathematician in the second half of the nineteenth-century.

LAGRANGE, Joseph Louis (1736–1813), French mathematician, made significant contributions to various mathematical disciplines such as number theory, theory of probability, theory of equations, analysis, calculus of variations, and also to mechanics.

LEIBNIZ, Gottfried Wilhelm (1646–1716), one of the leading mathematicians and philosophers of his time, is often referred to as the last scholar with universal knowledge. At the University of Leipzig (Germany), which he entered at the age of fifteen, Leibniz studied Theology and Law apart from Mathematics and Philosophy. Leibniz' chief contribution to mathematics was his discovery of calculus, at about the same time as, but independently of, Newton. Leibniz' interest in infinite processes, in particular in the study of infinite series, accounts for his fascination with the harmonic triangle displaying the sums of a variety of infinite series (see Appendix I (a)4).

LOBACHEVSKY, Nicolai Ivanovitch (1793–1856), eminent Russian mathematician, discovered the so-called hyperbolic geometry, independently of Bolyai. This was the first example of a geometry in which Euclid's parallel postulate was not valid.

MARKOFF, A.A. (1856–1922), Russian mathematician, worked in number theory.

MATIJASEVIČ, Jurii Vladimirovič, contemporary Soviet mathematician, solved one of Hilbert's celebrated 23 problems.

MINKOWSKI, Hermann (1864–1909), a mathematician born in Russia, studied and taught at universities in Germany and Switzerland. His interest in the application of geometric methods in solving number-theoretic problems led to the foundation of the geometry of numbers, a new branch of number theory. Minkowski is also celebrated for his work on providing mathematical descriptions of phenomena in the theory of relativity.

MOTZKIN, T.S., contemporary mathematician working in the U.S.A.

NEWTON, Isaac (1643–1727), English mathematician and scientist, is famous for his discovery of calculus and the universal law of gravitation. (Calculus was discovered also by Leibniz, in Germany, independently of Newton.) Newton's contributions to mathematics include the binomial theorem, and his contributions to physics his results on the nature of colours. Newton's *Principia* may be said to be one of the greatest achievements of the human mind. It was an eminently successful 'system of the world' which influenced all areas of human thought.

PAPPUS of Alexandria in about 320 A.D. composed a work entitled *Synagogue* (Collection), providing a detailed record of mathematical statements known at that time. New discoveries and generalizations are also included in the text.

PASCAL, Blaise (1623–62), a French mathematician prodigy. At the age of fourteen Pascal joined his father at meetings of celebrated mathematicians in Paris, and at the age of sixteen he published his famous essay on conics. Interest in problems on probability led Pascal to a detailed study of the so-called Pascal's triangle and he discovered many of its fascinating properties. In later life Pascal abandoned mathematics and science for theology. His philosophical writings are profound.

PEANO, Giuseppe (1858–1932), an Italian mathematician, interested in mathematical logic, is famous for his work on the foundations of arithmetic. Peano's mathematical work was distracted by his activity on the invention of an international language — a forerunner of Esperanto, which he called 'Interlingua', with words adapted from Latin, French, English and German.

PYTHAGORAS of Samos (circa 580–500 B.C.) was a philosopher and a mystic. His philosophical principles were based on properties of numbers. This being so, Pythagoras and his followers pursued mathematical studies. Although the secret society established by Pythagoras played an important role in cultivating mathematics, nothing is known about the mathematical discoveries of its members. Pythagoras' life is enmeshed in legend. The famous 'theorem of Pythagoras' was certainly known before his time. It is believed that the words 'philosophy' — that is 'love of wisdom' — and 'mathematics' — i.e. 'that which is learned' — were coined by him.

RAMSEY, Frank Plumpton (1903–30) worked as a lecturer in Mathematics at the University of Cambridge. He contributed notable results on the foundations of mathematics and also on mathematical economics; his vocation, however, was philosophy.

SCHOENBERG, Isaac, contemporary mathematician working in the U.S.A.

SIERPIŃSKI, Waclaw (1882–1969), Polish mathematician, made valuable contributions to analysis and the theory of numbers; his most important work concerns set theory and topology.

STEINHAUS, Hugo (1887–1972), Polish mathematician. Apart from his contributions to analysis, Steinhaus is well known for his interest in problems of elementary mathematics.

SYLVESTER, James Joseph (1814–97), British mathematician, a life-long friend of Arthur Cayley. He wrote important papers in number theory and analysis, but perhaps his favourite field of study was algebra. Together with Cayley he influenced the development of modern algebra. Sylvester was a stimulating teacher; his most celebrated pupil was a non-mathematician: the nurse Florence Nightingale.

WARING, Edward (1734–93) studied and taught at the University of Cambridge. Apart from a number of significant mathematical discoveries, Waring posed the following challenging question: Can every positive integer be written as a sum of a fixed number of kth powers of integers, for any given natural number k? Waring's question was answered affirmatively by Hilbert in 1909.

WILSON, John (1741–93), a friend and pupil of Waring, left mathematics for law.

Appendix III   Recommended reading (including references)

A. Puzzles and problems for solution

(a) Puzzles for beginners

[1] B.A. Kordemsky, *The Moscow Puzzles*, Penguin Books, 1978.
[2] Ya. Perelman, *Figures for Fun*, Mir Publ., Moscow, 1979.

(b) More puzzles and mathematical diversions

[3] H.E. Dudeney, *Amusements in Mathematics*, Dover, New York, 1958.
[4] M. Kraitchik, *Mathematical Recreations* (2nd edn), Dover, New York, 1953.
[5] J.S. Madachy, *Madachy's Mathematical Recreations*, Dover, New York, 1978.
[6] A. Dunn, *Mathematical Bafflers*, Dover, New York, 1980.
[7] C. Lukács and E. Tarján, *Mathematical Games*, Granada, London, 1982.

(c) Mathematical recreations, accompanied by essays and explanations of methods for solution

[8] W.W.R. Ball and H.S.M. Coxeter, *Mathematical Recreations and Essays*, (12th edn), Univ. of Toronto Press, 1974.
[9] E.P. Northrop, *Riddles in Mathematics*, Penguin Books, 1971.
[10] M. Gardner, *Mathematical Puzzles and Diversions*, Penguin Books, 1976.
[11] M. Gardner, *More Mathematical Puzzles and Diversions*, Penguin Books, 1976.
[12] M. Gardner, *Mathematical Circus*, Penguin Books, 1980.
[13] L.A. Graham, *Ingenious Mathematical Problems and Methods*, Dover, New York, 1959.

[14] T.H. O'Beirne, *Puzzles and Paradoxes; Fascinating Excursions in Recreational Mathematics*, Dover, New York, 1965.

(d) Suggestions for investigations and explanations of methods

[15] A. Gardiner, *Mathematical Puzzling*, Oxford Univ. Press, 1987.
[16] A. Gardiner, *Discovering Mathematics*, The Art of Investigation, Oxford Univ. Press, 1987.

(e) Competition problems

[17] C.T. Salkind and J.M. Earl, *The Contest Problem Book*, Parts I, II, III (annual high school contests in USA, from 1950 to 1972), New Mathematical Library, Vols. 5, 17 and 25. Published by the Mathematical Association of America.
[18] S.L. Greitzer, International Mathematical Olympiads 1959–77, New Mathematical Library, Vol. 27. Published by the Mathematical Association of America.
[19] Hungarian Problem Books I and II based on the Eötvös Competitions 1894–1905 and 1906–28. New Mathematical Library, Vols. 11, 12. Published by the Mathematical Association of America.
[20] D.O. Shklarsky, N.N. Chentzov, I.M. Yaglom. *The USSR Olympiad Problem Book*, Freeman, San Francisco and London, 1962.

(f) Challenging problems for very advanced pupils

[21] H. Steinhaus, *100 Problems in Elementary Mathematics*, Basic Books, New York, 1964.
[22] A.M. Yaglom and I.M. Yaglom, *Challenging Mathematical Problems with Elementary Solutions*, Holden Day, 1964.
[23] D.J. Newman, *A Problem Seminar*, Springer, New York, Heidelberg, Berlin, 1982.

B. Acquiring techniques for problem solving through the study of topics usually not covered in the secondary school syllabus

The following books in the series (a), (b), listed below, are highly recommended for advanced pupils interested in mathematics:

(a) Books in the series 'New Mathematical Library', published by the Mathematical Association of America. (The series began in 1961; the number of volume in the series is indicated in front of each title.)

[24] 1. I. Niven, *Numbers, Rational and Irrational*, 1961
[25] 2. W.W. Sawyer, *What is Calculus About?*, 1961
[26] 3. E.F. Beckenbach and R. Bellman, *An Introduction to Inequalities*, 1961
[27] 4. N.D. Kazarinoff, *Geometric inequalities*, 1961
[28] 6. P.J. Davis, *The Lore of Large Numbers*, 1961
[29] 7. L. Zippin, *Uses of Infinity*, 1962
[30] 8 and 21. I.M. Yaglom, *Geometric Transformations* I, II, 1962
[31] 9. C.D. Olds, *Continued Fractions*, 1963
[32] 10. O. Ore, *Graphs and Their Uses*, 1963
[33] 14. I. Grossman and W. Magnus, *Groups and Their Graphs*, 1964
[34] 15. I. Niven, *The Mathematics of Choice*, 1965
[35] 18. W.G. Chinn and N.E. Stewart, *First Concepts of Topology*, 1966
[36] 19. H.S.M. Coxeter and S.L. Greitzer, *Geometry Revisited*, 1967
[37] 20. O. Ore, *Invitation to Number Theory*, 1967
[38] 21. A. Sinkov, *Elementary Cryptanalysis — A Mathematical Approach*, 1966
[39] 22. R. Honsberger, *Ingenuity in Mathematics*, 1970
[40] 26. G. Pólya, *Mathematical Methods in Science*, 1977
[41] 28. E.W. Packel, *The Mathematics of Games and Gambling*, 1981

(b) Books in the series 'Little Mathematics Library', Mir Publications, Moscow.

[42] A.I. Markushevich, *Remarkable Curves*, 1980.
[43] L.I. Golovina and I.M. Yaglom, *Induction in Geometry*, 1979.
[44] V.A. Uspensky, *Pascal's Triangle and Certain Applications of Mechanics to Mathematics*, 1979.
[45] I. Lyubich and L.A. Shor, *The Kinematic Method in Geometrical Problems*, 1986.
[46] A.S. Solodovnikov, *Systems of Linear Inequalities*, 1979.
[47] Ye.S. Venttsel, *Elements in Game Theory*, 1980.

(c) More reading suitable for advanced pupils (highly recommended)

[48] N.N. Vorobyov, *The Fibonacci Numbers*, published in the series Topics in Maths, Heath, Boston, 1966.

[49] N. Vasiljev and V. Gutenmacher, *Straight Lines and Curves*, Mir Publ., Moscow, 1980.

[50] N. Vilenkin, *Combinatorial Mathematics for Recreation*, Mir Publ., Moscow, 1972.

[51] A.H. Beiler, *Recreations in the Theory of Numbers* (2nd edn), Dover, New York, 1966.

[52] N.N. Vorob'ev, *Criteria for Divisibility*, published in the series Popular Lectures, Univ. of Chicago Press, 1980.

[53] W. Sierpiński, *A Selection of Problems in the Theory of Numbers*, Macmillan, N.Y., 1964.

[54] I.Ya. Bakelman, *Inversions*, published in the series Popular Lectures, Univ. of Chicago Press, 1974.

[55] C.S. Ogilvy, *Excursions in Geometry*, Oxford University Press, 1976

[56] H.E. Huntley, *The Divine Proportion*, Dover, New York, 1970.

[57] D. Pedoe, *Circles, A Mathematical View*, Dover, New York, 1979.

[58] V.G. Boltyanski, *Equivalent and Equidecomposable Figures*, D.C. Heath, Boston, 1963.

[59] Four books in the series 'Dolciani Mathematical Expositions', published by the Mathematical Association of America, 1973–79:
Vol. I: R. Honsberger, *Mathematical Gems*.
Vol. II: R. Honsberger, *Mathematical Gems II*.
Vol. III: R. Honsberger, *Mathematical Morsels*.
Vol. IV: edited by R. Honsberger, *Mathematical Plums*.

(d) Books for pre-university pupils and for undergraduate university students

[60] G.H. Hardy and E.M. Wright, *An Introduction to the Theory of Numbers* (5th edn), Oxford Univ. Press, 1985.

[61] I.N. Herstein, *Topics in Algebra*, Blaisdell, 1964.

[62] K. Knopp, *Infinite Sequences and Series*, Dover, New York, 1956.

[63] D. Hilbert and S. Cohn Vossen, *Geometry and Imagination*, Chelsea, New York, 1952.

[64] V. Boltjansky and I. Gohberg, *Results and Problems in Combinatorial Geometry*, Cambridge Univ. Press, 1985.

[65] H.J. Ryser, *Combinatorial Mathematics* (2nd edn), The Carus Mathem. Monographs No. 14, published by the Mathematical Association of America, 1965.

[66] S.S. Anderson, *Graph Theory and Finite Mathematics*, Markham Publ., Chicago, 1970.

[67] R.J. Wilson, *Introduction to Graph Theory* (2nd edn), Longman, Harlow, 1981.

[68] H. Steinhaus, *Mathematical Snapshots*, Oxford Univ. Press, 1960.
[69] I.J. Schoenberg, *Mathematical Time Exposures*, The Mathematical Association of America, 1982.
[70] P.G. Doyle and J.L. Snell, *Random Walks and Electric Networks*, The Carus Mathem. Monographs, No. 22, 1984.

(e) Polya's classical book of hints for problem solvers

[71] G. Polya, *How to Solve it*, Doubleday Anchor Books, New York Garden City, New York, 1957.

C. Excursions into the history of mathematics

[72] A. Aaboe, *Episodes from the Early History of Mathematics*, New Mathematical Library, Vol. 13, published by the Mathematical Association of America, 1964.
[73] K.O. Friedrichs, *From Pythagoras to Einstein*, New Mathematical Library, Vol. 16, published by the Mathematical Association of America, 1965.
[74] H. Dörrie, *100 Great Problems of Elementary Mathematics*, Dover, New York, 1965.
[75] V.G. Boltianski, *Hilbert's Third Problem*, Wiley, New York, 1978.
[76] H. Eves, Great *Moments in the History of Mathematics*, Dolciani Mathematical Expositions Nos. 5 and 7, published by the Mathematical Association of America, 1981.
[77] F. Klein *et al., Famous Problems*, Chelsea, New York, 1955.
[78] P. Dedron and J. Itard, *Mathematics and Mathematicians*, Vols. 1, 2. The Open Univ. Press, Milton Keynes, 1978.
[79] H. Eves, An Introduction to the History of Mathematics, Holt, Rinehart and Winston, New York, 1964.
[80] C.B. Boyer, A History of Mathematics, Wiley, 1968.
[81] H.L. Resnikoff and R.O. Wells, Jr., Mathematics in Civilization, Dover, New York, 1984.

D. What is mathematics? (especially recommended for pre-university pupils)

[82] R. Courant and H. Robbins, *What is Mathematics?* Oxford Univ. Press, 1978.

[83] A.D. Aleksandrov, A.N. Kolmogorov, M.A. Lavrent'ev (eds.), *Mathematics, Its Content, Methods and Meaning* (3rd edn), M.I.T. Press, Cambridge, Massachusetts, 1981.

[84] H. Rademacher and Toeplitz, *The Enjoyment of Mathematics* (2nd edn), Princeton Univ. Press, 1970.

[85] H. Rademacher, *Higher Mathematics from an Elementary Point of View*, Birkhauser, Boston, Basel, Stuttgart, 1983.

[86] I. Stewart, *The Problems of Mathematics*, Oxford Univ. Press, 1987.

[87] C.S. Ogilvy, *Tomorrow's Maths* (2nd edn), Oxford Univ. Press, 1972.

[88] H. Meschkowski, *Unsolved and Unsolvable Problems in Geometry*, Oliver & Boyd, Edinburgh, 1966.

[89] J. Lighthill *et al.*, *New Uses of Mathematics*, Penguin Books, 1980.

E. Further references

(a) Books

[90] H. Frasch and K.R. Löffler, *Bundeswettbewerb Mathematik — Aufgaben und Lösungen* 1972–78, Klett, Stuttgart, 1979.

[91] V. Sierpiński, *Co wiemy, a czego nie wiemy o liczbach pierwszych*, Państwowe Zaklady Wydawnictwo Szkolnych, 1961.

[92] W. Sierpiński, *Elementary Theory of Numbers*, Monografie Matematyczne, Tom 42, Warszawa, 1964.

[93] M. Hall, Jr. *Combinatorial Theory*, Blaisdell, Waltham, Massachusetts, Toronto, London, 1967.

(b) Journals

[94] A. Engel and H. Severin, *Internationale Mathematik-Olympiade*, Der Mathematikunterricht, Vol. 25, No. 1, 1979 (German).

[95] *Középiskolai Matematikai Lapok (Fizika Rovattal Bövitve) új sorozat*, Publisher: A tankönyvkiadó Vállalat, Budapest, Hungary (Hungarian).

[96] Kvant, *Naucho-populyarnij fiziko-matematicheskij zhurnal*. Publisher: Nauka, Moskova, USSR (Russian).

Index